POLLUTION SCIENCE, TECHNOLOGY AND ABATEMENT

RADIATION EXPOSURE

SOURCES, IMPACTS AND REDUCTION STRATEGIES

POLLUTION SCIENCE, TECHNOLOGY AND ABATEMENT

Additional books in this series can be found on Nova's website under the Series tab.

Additional E-books in this series can be found on Nova's website under the E-book tab.

ENVIRONMENTAL HEALTH - PHYSICAL, CHEMICAL AND BIOLOGICAL FACTOR

Additional books in this series can be found on Nova's website under the Series tab.

Additional E-books in this series can be found on Nova's website under the E-book tab.

POLLUTION SCIENCE, TECHNOLOGY AND ABATEMENT

RADIATION EXPOSURE

SOURCES, IMPACTS AND REDUCTION STRATEGIES

DAMIJAN BALENOVIC
AND
EMILIJE STIMAC
EDITORS

Nova Science Publishers, Inc.
New York

For permission to use material from this book please contact us:
Telephone 631-231-7269; Fax 631-231-8175
Web Site: http://www.novapublishers.com

Additional color graphics may be available in the e-book version of this book.

Library of Congress Cataloging-in-Publication Data

Radiation exposure : sources, impacts, and reduction strategies / editors, Damijan Balenovic and Emilije Stimac.
p. ; cm.
Includes bibliographical references and index.
ISBN 978-1-62100-869-9 (hardcover)
I. Balenovic, Damijan. II. Stimac, Emilije.
[DNLM: 1. Environmental Exposure. 2. Radiation Effects. 3. Radiation Protection. WN 600]

363.17'99--dc23

2011039237

Published by Nova Science Publishers, Inc. † New York

CONTENTS

PREFACE

Soil is the main source of continuous radiation exposure to humans. It also acts as a medium of migration for transfer of radionuclides to the biological systems and hence, it is the basic indicator of radiological contamination in the environment. In this new book, the authors present current research in the study of the sources, impacts and reduction strategies to radiation exposure. Topics discussed include radiation exposure and protection in conventional pediatric radiology of the thorax; natural and man-made radionuclides in the soil; the current state of terrestrial ecosystems in the Eastern Ural radioactive trace; pharmaceuticals photochemical reactions and radiation exposure in pregnancy and in medical imaging.

Chapter 1 - Soil is the main source of continuous radiation exposure to human. It also acts as a medium of migration for transfer of radionuclides to the biological systems and hence, it is the basic indicator of radiological contamination in the environment. Radionuclides in soil take part in several biogeochemical processes that eventually determine their mobilization and availability for ecological processes. Knowledge of their distribution in soil is very important in radiation protection research because it forms a baseline for impact assessments in the environment wherever there is radioactive contamination from practices or releases from accidents. The soil radioactivity is important for the purposes of establishing baseline data for future radiation impact assessment, radiation protection and exploration.

Natural environmental radioactivity and the associated external exposure due to gamma radiation depend primarily on the geological and geographical conditions, and appear at different levels in the soils of each region in the world. In addition to natural radionuclides, radiocaesium is a radiologically important contaminant of the soil environment. Radiocaesium was introduced into the atmosphere through nuclear weapons testing and nuclear accident especially that in Chernobyl. Due to its long physical half-life, high fission yield and very low migration velocity in soil environment, this radionuclide is very hazardous in the radioecological context. In many soil conditions it is readily available for uptake to plants resulting in potential contamination of the human food chain.

In this chapter the characteristics of radionuclides commonly present in soil and factors influencing their distribution in soil are summarized. The results of extensive studies conducted in Serbia on spatial distribution of radionuclide activity concentrations in soil, their mutual relationships and assessments of radiation exposure of population are presented.

Chapter 2 - Radiation protection depends on many factors that can be divided on the ground of technical and human factors that are both equally relevant for the radiation protection. To provide the best care for the young patients the effects of radiation should be minimized in accordance with ALARA (As Low As Reasonably Achievable) principle. Chest radiograms represent the basic radiological examination of thorax as they provide diagnostic information on the lung and heart status, and the skeletal system. To provide the best care for the young patients the effects of ionizing radiation should be minimized. Dosimetry is an indispensable part of every investigation related to the optimization and improvement of radiation protection. This study is dealing with the significance of different dosimetry systems that are applied in X-rays diagnostics, determination of doses received by patients and with the role of human factor during the radiological examination. Also the impact of ionizing radiation on cytogenetic status of children exposed to X-rays, since understanding the risks of low doses of radiation is an important aspect in the risk benefit analysis in pediatric populations. The study comprised child population that had a clinical indication for chest X-ray examination. Absorbed dose measurements were conducted on the forehead, thyroid gland, gonads, chest and back of the patients as well as on the phantom for radiological examinations of thorax. In this work doses were measured using various thermoluminescence (TL) and recently radiophotoluminescence (RPL) dosimetry systems. RPL and TL dosimeters were placed on the patients at the entrance of the beam in the centre of the X-ray field to measure the entrance dose. For comparison, during chest radiography, the dose in anthropomorphic child phantom was measured with TL dosimeters. The results on patients with both dosimetry systems showed a good correlation of measured dose. Also dose values on phantom were comparable with the results obtained on patients. If all technical malfunctions are excluded equal responsibility is on radiology technicians that are responsible for the patient's dose. Depending on their education and experience, technicians perform X-ray examination with various end results. In this study doses measured by two technicians were in a good agreement, for the third one doses were significantly higher. Following the analysis of the dosimetry results and the education of the personal the doses measured on children for all of them were in very good agreement and lower than before education. For the cytogenetic monitoring comet assay and micronucleus test were applied as a sensitive biomarkers for the detection of DNA damage and chromosomal instability after irradiation. Results demonstrated that there is significant potential of inducing genetic instability with even the low level diagnostic X-ray procedures. Radiological examination of children is extremely significant since it stands for the sole and most accurate investigation method. Radiation protection depends on many factors and a suitable cooperation of all medical doctors especially radiologists would be recommended. In conclusion, this work shows the importance of continues education and good teamwork for dose reduction to minimize the potential damage and maintain the optimal benefit for the young patients.

Chapter 3 - This Chapter focuses on the analysis of data resulting from almost 60 years of studies aimed at assessment of the hemopoiesis status in Techa riverside residents exposed to low-dose rate (LDR) radiation in the range from low to high doses. The red bone marrow (RBM) doses (individual values: 0.01-6.69 Gy; mean value - 0.89 Gy) were due to intakes of ^{90}Sr in drinking water and local food products, and also to external γ-exposure. The status of hemopoiesis (cellular composition of the RBM and peripheral blood, incidence of leukemia and RBM hypoplasia) was monitored from 1951 through 2008. Peripheral blood cell counts performed for exposed riverside residents during the period of maximum radiation exposures (1951-1953) displayed a significant ($p<0.0001$) decrease in the proportion of leukocytes (neutrophils and lymphocytes) and thrombocytes. Peripheral blood erythrocyte count was maintained at the normal level, owing to the efficacy of compensatory processes in the bone marrow, including increased rates of erythrokaryocyte proliferation and maturation. The development of peripheral blood granulocytopenia resulted from a delay in the differentiation of neutrophilic bone-marrow granulocytes, a marked increase in the frequency of lethal abnormalities in bone-marrow neutrophils, and pathological mitoses. The periods of recovery of the blood cell composition differed considerably between different blood cell series, and depended on exposure dose rates, degree of the primary inhibition of hemopoiesis, as well as on individual physiological characteristics of the exposed persons (gender, age at initial exposure). At late time, an increased risk of leukemia was registered.

Chapter 4 - A large-scale nuclear accident took place at the Production Association "Mayak" in the Southern Urals in 1957. About 74×10^{15} Bq of radioactive substances were released into the atmosphere, which resulted in the contamination of a vast area, the Eastern-Ural Radioactive Trace (EURT). In this case, the main long-lived radionuclide was ^{90}Sr. In 1967, this area was subjected to a secondary contamination resulting from radioactive sediments transported by wind from the Mayak technological reservoir, Karachay Lake. The ^{137}Cs was predominant in the composition of this contamination. Two contamination range areas were established: impact and buffer zones. Within EURT area the radionuclide contamination density decreases with the distance from the accident location and is fairly described as a power function. The specific activity of radionuclides in the herbaceous plants depend on the density of soil contamination. However, transfers coefficients for plants of impact zone were, as a rule, lower than that for background one. Calculations indicated that inpresent stock of the radionuclides in EURT soils is 570×10^{12} Bq for ^{90}Sr, 66×10^{12} Bq for ^{137}Cs and about 2×10^{12} Bq for ^{239,240}Pu. The stock in the impact zone consist of 75% of ^{90}Sr, 70% of ^{239,240}Pu and only 23% of ^{137}Cs from their total amount within EURT zone. Most of ^{137}Cs stock is located in the buffer area where it was received as a result of the Karachay accident. Doze loads upon some herbaceous plants were evaluated. In the impact zone, they exceed background level on 1–3 orders of magnitudes. The aspects of flora diversity within the EURT were studied. The natural growth in the EURT zone is a complex of synanthropic and semi-natural cenosises on the different stages of degradation and restoration. Species diversity in the impact zone is very high; some communities have up to 65 species, it is associated with a low anthropogenic load. The current state of phytocenosises in the demolished local villages is still largely determined by the degree of man's impact in the pre-accident period and recultivation activities. The natural ecosystems successfully recovered at a faster pace.

Chapter 5 - *Introduction*: This chapter focuses on the topic of Radiation exposure during Pregnancy. The indications for performing radiological procedures to aid in the clinical management of the pregnant patient and the appropriate triaging of the diagnostic modalities are discussed. A brief description of the salient features of Radiation biology as well as effects of radiation on the fetus and the mother is presented. Finally dose reduction strategies available to minimize risk to the fetus, and the need for and the process of obtaining informed consent and counseling the pregnant patient are discussed.

Clinical Indications and Imaging Algorithms: Common indications for performing radiological procedures during pregnancy include diagnosis of Pulmonary Embolism, evaluation of acute abdominal pain for conditions such as acute appendicitis, urolithiasis, and biliary colic as well as assessment of traumatic injuries in a pregnant patient. Fluoroscopy for guidance of interventional procedures and the very rare need for radiation treatment of cancer in pregnancy are additional sources of medical needed radiation exposure in pregnancy.

Radiation Biology, Fetal Radiation Risks, Maternal Radiation Risks: There are threshold effects resulting from exposure to radiation in utero and these include birth defects, mental retardation, neurobehavioral effects, and growth retardation and embryonic death. Radiation induced carcinogenesis is a non threshold phenomena. An example of radiation risk to the mother includes an increased risk for breast cancer resulting from Chest CT performed for diagnosis of pulmonary embolism.

Radiation Dose Reduction Strategies: There are various strategies that can be put in place to reduce radiation dose resulting from Computed Tomographic examinations. Fluoroscopy performed in conjunction with urgent and emergent interventional procedures is an important source of radiation and although less often utilized compared to CT, it is important to adopt techniques to keep the radiation dose as low as possible during fluoroscopy.

Counseling and Consent: The American College of Radiology recommends that written informed consent be obtained in all instances when a pregnant woman is exposed to medically indicated radiation. Counseling of the patient of the risks and the benefits of the procedure is an important component of the consent process. Likewise referring clinicians also need to be educated on the potential hazards and appropriateness of use of radiation during pregnancy.

Conclusion: Exposure to radiation in pregnancy may be medically indicated for an accurate and timely clinical diagnosis, to guide interventional procedures and very rarely to treat cancers. Medical necessity in each instance must outweigh the potential risk to the fetus. Whenever feasible, exposure to radiation must be deferred, replaced by methods that do not use ionizing radiation or when used, must utilize the lowest possible dose so as to keep the deleterious effects to the fetus and the mother at a minimum.

Chapter 6 - Photolytic degradation can be a major limiting factor in the stability of pharmaceuticals. A drug may be chemically affected by the radiation of a particular wavelength only if it absorbs radiation at that wavelength and if the power exceeds a threshold. Ultraviolet radiation, which has a high energetic level, is the cause of many degradation reactions. If the molecule absorbs radiation, it produces excited states of itself and may react photochemically with other species. When absorbing molecules are not directly involved in the reaction, but they transfer their energy to other reactive molecules, the absorbing substance is called photosensitizer. Due to the fact that many experimental variables may be involved in photochemical reaction, the kinetic may be very complex. The intensity and wavelength of light and the size, shape, composition and color of the container can affect the reaction rate. Photodegradation of some pharmaceuticals follows zero order kinetics; moreover, others are degraded following a kinetic of first order.

This chapter includes:

a. The use of UV radiation for degradation of pharmaceuticals present in industrial waste.
b. The use of UV radiation to catalyze the reaction of omeprazole degradation, and their determination based on the modification in the fluorescent properties.

Chapter 7 - The terrestrially available UVB was measured underwater on a fringing coral reef at the southern end of Australia's Great Barrier Reef marine park (25°S, 153°E) using poly 2,6-dimethyl-1,4-phenylene oxide (PPO) film dosimeters. These dosimeters have been employed in previous research applications to measure long term terrestrial UV exposures on land and in controlled aquatic environments (typically for exposures lasting several days). The value of these dosimeters is discussed in reference to their use in an underwater coral reef environment extending over periods of several weeks. The UVB measured in the autumn at the test reef site indicates that exposures in a shallow coral reef located at a sub-tropical latitude vary from between 34 kJm^{-2} and 50 kJm^{-2} over a period of 28 days equating to a mean daily UVB exposure of between 1 and 2 kJm^{-2}. These measurements made at an approximate depth of 2 m represent approximately 5% of the daily surface UVB. The results indicate that the film dosimeters used have the potential to be applied in a number of different underwater locations to investigate the impact of solar UV radiation on exposed coral surfaces.

Chapter 8 - The last decades have seen an unprecedented increase in technology in medicine. Among this development, imaging has taken one of the leading positions. Nevertheless, this increase in medical imaging use comes at a significant cost. The most common medical imaging techniques used radiation to acquire the image. From the initial X ray technology form the end of the nineteen century to today's computed tomography and nuclear medicine technology, medical image came from an insignificant to the second most important cause of radiation exposure throughout the world.

Although this information may sound negative, medical imaging has an invaluable use in diagnostic and therapeutic medicine. Therefore, it should not be blamed for the radiation exposure; it should only be controlled and adequately used to optimize the medical benefits while reducing the risks associated with the increase in radiation exposure.

In this chapter the authors will discuss the increase in radiation exposure from medical imaging, while focusing on potential technical interventions that can help reducing exposure and optimize equipment to prevent the use of unnecessary radiation.

Chapter 9 - In recent years, there has been a greater concern with the effects of ultraviolet (UV) radiation in pharmaceutical products such as drugs and health materials. The photostability is an important parameter to evaluate the demanded quality, security and effectiveness for the safe use of health products. Due to environmental conditions in which the product will be exposed, such as temperature, humidity and light during its marketing period, photostability tests are important to monitor the degradation reactions and to predict shelf life of the final product in normal storage conditions. The sun is the most important source of UV radiation, although visible and infrared light constitute the major fraction of the radiation reaching Earth´s atmosphere. Artificial lights are also sources of UV radiation. Incandescent lamps, halogen lamps and fluorescent tubes are the most commonly used light sources in homes and work places. Incandescent lamps give practically no UV radiation, while halogen lamps operate at a higher temperature and give small fluence rates of UVA radiation. Fluorescent tubes contain mercury vapor which give a typical band spectrum showing two strong lines located at the UV region (λ = 254 and 313 nm) which are not completely filtered by the glass tube. Photochemical degradation can be an important factor in the stability of pharmaceuticals especially under exposure of high energy radiation such as the UV radiation. So, not only the drug photostability evaluation is an important step in this process, but also it is important to know and to quantify (i) the formed byproducts, (ii) the degradation kinetics and (iii) the influence of light on the original formulation. It is usually difficult to predict the light effect in the compounds and the consequences to the stability of the final preparation. So, to perform photostability testing, it is necessary to ensure that the radiation is homogeneously distributed on the surface area where the product will be exposed and to precisely determine the radiation dose received by the samples when placed in the exposure area. Thereby, it is necessary to design and to qualify a photostability chamber suitable to perform photostability testing and to guarantee the reliability of the results provided by it. Either a commercial or a home-built photostability chamber needs to ensure homogeneous light dose on the studied sample. There are two experimental ways to qualify the source of radiation: (i) by applying actinometer systems (chemical compounds with a linear photodegradation rate) or (ii) by using a physical device on which the number of photons applied can be fully determined. The main purpose of this Chapter is to discuss the procedures involved in the degradation induced by UV radiation and how to monitore and to determine the correct conditions for performing photostability testing in a photostability chamber.

In: Radiation Exposure
Editors: D. Balenovic and E. Stimac
ISBN 978-1-62100-869-9

Chapter 1

Natural and Man-Made Radionuclides in Soil as Sources of Radiation Exposure

Snežana D. Dragović[a], Ljiljana J. Janković-Mandić[b], Ranko M. Dragović[c], and Milan M. Đorđević[c]

[a]University of Belgrade, Institute for the Application of Nuclear Energy, Banatska 31b, 11080 Belgrade, Serbia

[b]UniversityofBelgrade,Vinča Institute of Nuclear Sciences, P.O. Box 522, 11001 Belgrade, Serbia

[c]University of Niš, Faculty of Science and Mathematics, Department of Geography, Višegradska 33, 18000 Niš, Serbia

Abstract

Soil is the main source of continuous radiation exposure to human. It also acts as a medium of migration for transfer of radionuclides to the biological systems and hence, it is the basic indicator of radiological contamination in the environment. Radionuclides in soil take part in several biogeochemical processes that eventually determine their mobilization and availability for ecological processes. Knowledge of their distribution in soil is very important in radiation protection research because it forms a baseline for impact assessments in the environment wherever there is radioactive contamination from practices or releases from accidents. The soil radioactivity is important for the purposes of establishing baseline data for future radiation impact assessment, radiation protection and exploration.

Natural environmental radioactivity and the associated external exposure due to gamma radiation depend primarily on the geological and geographical conditions, and appear at different levels in the soils of each region in the world. In addition to natural radionuclides, radiocaesium is a radiologically important contaminant of the soil environment. Radiocaesium was introduced into the atmosphere through nuclear weapons testing and nuclear accident especially that in Chernobyl. Due to its long physical half-life, high fission yield and very low migration velocity in soil environment, this radionuclide is very hazardous in the radioecological context. In many soil conditions it is readily available for uptake to plants resulting in potential contamination of the human food chain.

In this chapter the characteristics of radionuclides commonly present in soil and factors influencing their distribution in soil are summarized. The results of extensive studies conducted in Serbia on spatial distribution of radionuclide activity concentrations in soil, their mutual relationships and assessments of radiation exposure of population are presented.

1. Introduction

Defining and understanding radionuclide distribution in soils is a key issue in many environmental studies. Soils play a major role in biogeochemical cycles of elements and nuclides. Elements are contained within the crystal structure of soil-forming minerals as well as bound to different phases of soil particles by a variety of mechanisms. Physico-chemical soil properties influence the element mobility and their bioavailability in the terrestrial ecosystems (Kabata-Pendias and Pendias, 2005). The geochemical composition of soils is of primary importance in establishing environmental baselines of both essential and potentially toxic elements. Soil is the main reservoir for long-lived radionuclides deposited on terrestrial ecosystems via different pathways: as a part of Earth's original crust (primordial radionuclides), produced and deposited by cosmic ray interactions (cosmogenic radionuclides) or as a result of anthropogenic activities (man-made radionuclides) (UNSCEAR, 2010). Radionuclides in soils occur in minerals or are adsorbed onto soil components (organic matter, carbonates, iron and manganese oxides) and from an environmental point of view, processes acting on their transfer and bioavailability are an important issue.

The immobility of radioactive elements in uppermost soil layers represents a problem for the environment and human health, since they can be easily integrated in the food chain (Napier and Reed, 2006). The major part of radionuclides released into the environment is finally accumulated in either the upper layer of soils or interstitial system of sediments in aquatic systems. As a consequence, a risk for ecosystems and health could be significant.

2. Sources of Radionuclides in Soil

Primordial radionuclides are the naturally occurring forms of radioactive elements incorporated into the Earth during its formation. The most significant terrestrial radionuclides include the uranium and thorium decay series, ^{40}K and ^{87}Rb. Cosmogenic radionuclides are continuously produced from collisions between the highly energetic cosmic rays with stable nuclides in the atmosphere. Cosmogenic radionuclides include ^{3}H, ^{7}Be, ^{14}C and ^{22}Na. Man-made radionuclides are produced through anthropogenic activities, such as atmospheric testing of nuclear weapons and operation of nuclear facilities. Examples of man-made radionuclides that could be in the environment are ^{137}Cs, ^{90}Sr and various isotopes of plutonium.

The main contribution to external exposure comes from gamma-emitting radionuclides present in trace amounts in the soil, mainly ^{40}K and the ^{238}U and ^{232}Th decay series commonly associated in nature (UNSCEAR, 2010; Ivanovich, 1994). Numerous studies have been conducted to assess the radiation exposure due to these radionuclides and the results obtained are exploited to enrich the world's data bank, greatly needed for evaluating worldwide average values of radiometric and dosimetric quantities. According to UNSCEAR (2010) world average soil concentrations of ^{40}K, ^{238}U, ^{226}Ra and ^{232}Th are: 412, 33, 32 and 45 Bq kg^{-1}, respectively. The activity concentrations of these radionuclides in different regions of the world are listed in Table 1.

Activities related to exploitation of natural resources to meet the human needs can lead to enhanced levels of naturally occurring radioactive material in products, by-products and wastes.

Table 1. Natural radionuclides activity concentrations in soils worldwide

Region	^{40}K		^{238}U		^{226}Ra		^{232}Th	
	Mean	**Range**	**Mean**	**Range**	**Mean**	**Range**	**Mean**	**Range**
Africa	247	0-1150	21	1-120	25	5-180	16	1-140
North America	307	100-700	35	4-140	40	8-160	35	4-130
Central America	234	6-2260	46	11-130	34	0.5-115	8	0.05-42
South America	654	559-773	-	-	-	-	-	-
East Asia	436	7-2168	50	2-690	40	2-440	52	1-360
West Asia	365	4-780	28	6-70	26	0.5-97	26	1.5-60
North Europe	522	40-1200	37	10-1000	32	2-1000	27	0.5-115
West Europe	548	0-3200	38	2-543	40	5-310	38	1-258
East Europe	462	15-1599	29	0-190	35	1-275	33	2-168
South Europe	333	0-1570	42	2.5-190	28	0-310	26	0.8-193
Oceania	370	<40-740	-	-	23	<4-56	29	<4-63

UNSCEAR, 2010.

There are two effects of human exploitation that are relevant in the case of potential effects of naturally occurring radioactive material on human health and the environment: (1) the concentration of naturally occurring radioactive material can be enhanced above its natural levels in a product, by-product or residue; (2) the availability for release into the biosphere of the naturally occurring radioactive material in products, by-products or residues can be enhanced through physicochemical changes or by method by which the residues are managed (IAEA, 2003). Processes enhancing concentrations of naturally occurring radioactive material include: oil and gas production, coal mining and combustion, geothermal energy production, drinking water and wastewater treatment, production of industrial minerals, use of phosphate fertilizers and irrigation.

During the extraction of oil and natural gas, elevated activities of ^{226}Ra and ^{228}Ra present in underground formations are often released (UNSCEAR, 2010). The process residues associated with the oil and gas production occur in the form of scales, sludges and films. There are also additional naturally occurring radioactive material containing residues or contaminated soils from the water discharges produced. The largest concern in terms of radionuclide activity concentrations in the oil and gas industry involves the hard scales which form on the inside of the downhole tubing (IAEA, 2003). At many oilfield sites the downhole tubing can become coated with scale deposits that contain radium which, when removed, become a waste management concern. In terms of exposure, the scales tend to be relatively insoluble, thus the radionuclides would only be released slowly into the environment (Raabe, 1996). The disposal of scale from oil extraction industry installations and of sludge containing naturally occurring radioactive material can be of environmental significance, with contamination of land being the major concern. The average radium concentration in soils sampled at an oilfield contaminated with naturally occurring radioactive material in eastern Kentucky, USA, was 32,560 Bq kg^{-1}(Rajaretnam and Spitz, 2000).

Fossil fuels such as coal contain naturally occurring radionuclides from the uranium and thorium series and ^{40}K (Table 2). Most of the waste from a typical coal fired power plant is generated as fly ash, which is entrained with the hot flue gases of the combustion process. The remainder of the ash is heavier and settles to the bottom of the boiler to form bottom ash. The concentration of naturally occurring radioactive material in the coal, waste rocks and waste waters depends on the geological formation of coal seams.

Table 2. Activity concentrations of some natural radionuclides in coal (Bq kg^{-1})

Country	^{226}Ra	^{232}Th	^{238}U	^{40}K	Reference
Australia	19-24	11-69	8.5-47	23-140	Smith, 1987
Brazil	72	62	72	-	Clarke, 1993
Egypt	26	8	59	-	WISMUT, 1992
Germany	32	21	-	225	Borio et al., 1985
Greece	44-206	21	117-390	-	Martin et al., 1997
Hungary	-	12-97	20-480	30-384	Vandenhove et al., 2000
Italy	-	18	23	218	Borio et al., 1985
Poland	-	<123	<159	<785	Wysocka and Skowronek, 1991
Romania	126	<170	<415	-	Martin et al., 1997
United Kingdom	7.8-21.8	7-19	7-19	55-314	Smith, 1987
United States	8.9-59.2	3.7-21.1	6.3-73	-	Smith, 1987

The concentration of naturally occurring radioactive material in ash and other residues is closely related to the naturally occurring radioactive material concentrations in the coal used as fuel (IAEA, 2003). Resuspension of fly ash and other flue gas residues from open disposal pits could be relevant pathways for exposure. In addition to coal burning power plants emissions, particulate emissions from coal based domestic heating systems and the potential for exposure can be considerable, in particular in less developed countries (Clarke, 1992). During the burning of coal, the organic compounds are converted into gases (water vapour and carbon dioxide), while the inorganic elements, which include the significant naturally occurring radionuclides, are concentrated in the ashes (SCK-CEN, 2003). In general, the radionuclide enhancement factor is about 10. Papastefanou et al. (1988) reported enrichment of radium in soils around CPPs in Greece (up to 971 Bq kg^{-1}). The elevated ^{226}Ra activity concentrations (up to 337 Bq kg^{-1}) in soils around lignite field basin in southern Greece have been found by Rouni et al. (2001). Considerably elevated concentrations of ^{238}U (up to 944 Bq kg^{-1}) and ^{226}Ra (up to 883 Bq kg^{-1}) have been found in soil around coal-fired power plant in Hungary (Papp et al., 2002).

Naturally occurring radioactive material is found in the solid residues generated by the exploration and development of geothermal systems and in the extraction of the Earth's geothermal energy for use in either producing electric power or supplying direct heat. The wastes associated with geothermal energy production are similar to those associated with oil and gas production: temperature changes lead to precipitation of solids from hot formation waters in piping, equipment, and retention ponds at the surface. Residues produced contain significant concentrations of radium and its decay products (IAEA, 2003).

Conventional water treatment processes designed to remove suspended solids and dissolved chemical contaminants from drinking water supplies also remove radionuclides. Examples of naturally occurring radioactive material containing residues from municipal water treatment are radioactive contaminated sludges and solids including filter sludges, spent ion exchange resins, spent granular activated carbon, as well as waters from filter backwash (IAEA, 2003). Disposal of sludges in lagoons results in the accumulation of radium in the bottom sediments. Naturally occurring radioactive material containing sludge is also disposed in sanitary landfills, discharged to sewers, injected in deep wells or spread on agricultural soils as a fertilizer, while the decanted water is recycled. The soil contamination due to the spread of sludge depends on the radionuclide concentration in the sludge and on the thickness of the sludge layer on land (UNSCEAR, 2010).

The common practice of dispersing sewage sludges or incinerator ashes onto agricultural land can lead to the accumulation of radionuclides and heavy metals in the soils and their uptake by plants (UNSCEAR, 2010).

Several industries, such as pulp and paper industries, are heavy users of process water. Owing to the substances involved, significant radionuclide concentrations can accumulate in various products, intermediates and wastes (HPS, 2000).

As with drinking water, surface and groundwaters used for irrigation may contain naturally occurring radioactive material that may accumulate in certain soil fractions and eventually transfer to plants. An active leaching of uranium from marine shales has been observed in irrigated areas in southeastern Colorado (USGS, 1994).

Naturally occurring radioactive material containing residues are generated by the mining and processing of phosphate rock (phosphorite) that is processed to phosphoric acid or elemental phosphorus (IAEA, 2003). Uranium may be incorporated in sedimentary phosphorite ores through ionic substitution into the carbonate-fluorapatitic crystals or by adsorption. Igneous phosphorite contains less uranium, but more thorium. High phosphate contents usually correspond to high uranium contents. The significant enrichment of ^{238}U in soils (up to 984 Bq kg^{-1}) around old phosphate mine in Jordan has been reported by Al-Jundi (2002). The activity concentrations of ^{226}Ra up to 820 Bq kg^{-1} in soils around phosphate mines in Syrian Arab Republic (Othman and Al-Masri, 2007).

Releases of radioactivity associated with uranium mining and milling operations are considered to be a significant cause of environmental and radiological problems in and around mining facilities. Worldwide, the total estimated volume of tailings is 938 million cubic meters produced at more than 4000 mines (IAEA, 2004). Improper disposal of mill tailings can lead to substantial contamination of soils, surface water and groundwater. In addition to radioactivity, uranium mill tailings are often associated with elevated concentrations of highly toxic heavy metals. The ^{238}U and ^{232}Th activity concentrations in soils of uranium mining and milling areas wordwide are given in Table 3.

Table 3. The ^{238}U and ^{232}Th activity concentrations (Bq kg^{-1}) in soils of uranium mining and milling areas wordwide

Country	^{238}U	^{232}Th	Reference
Australia	48-6780	-	Hancock et al., 2006
Brazil	41.6-180	74.0-99.0	Fernandes et al., 2006
Kyrgyzstan	49-1593	-	Vandenhove et al., 2006
Namibia	53-1752	70.4-1866	Oyedele et al., 2010
Portugal	123-337	-	Carvalho et al., 2007
Portugal	392	-	Carvalho et al., 2009
Spain	6043	-	Blanco et al., 2005
United States	20-2320	-	Rood et al., 2008

Major sources of man-made radionuclides in soil, and generally in the environment, are the production and testing of nuclear weapons, nuclear power plants, nuclear fuel reprocessing, repository of nuclear wastes and nuclear accidents. High-level waste liquid and sludge are generated from weapons production and reprocessing, by which fissile materials are separated from fission products in spent nuclear reactor fuels to produce weapons-grade fuel. Nuclear weapons testing began in New Mexico in 1945 and since then over 2×10^{8} TBq of radioacivity has been released into the atmosphere (Choppin, 2003). In the United States a total of 1054 nuclear tests have been conducted until 1992, taking place in the Pacific Ocean, at the Nevada test site and on miscellaenous sites in the continental United States (USDOE/NV, 1992). Testing at Nevada test site contributed to the global fallout budget as well as produced a legacy of radioactivity on the surface and below ground. Very low $^{137}Cs/^{239,240}Pu$ ratios and high $^{239,240}/^{238}Pu$ ratios, in addition to high activity concentrations of $^{239,240}Pu$ in soils around Nevada test site, reported by Turner et al. (2003), have been attributed to nuclear weapons tests. During 1949-1989, the Soviet Union conducted a total of about 460 nuclear weapons tests in Semipalatinsk, including 133 atmospheric nuclear tests. Starting in 1961, more than 300 tests explosions were conducted underground. Thirteen of the underground tests resulted in releases of radioactive gases to the atmosphere (IAEA, 1998). In late 1940s to early 1960s, about 350 atmospheric atomic bomb tests were conducted at different test sites in the northern hemisphere (Eikenberg et al., 2004). Between 1966 and 1996, France had conducted nuclear weapons tests in French Polynesia (Pfingsten et al., 2001), where particles containing plutonium and small amounts of americium resulting from safety trials are found in soils and sediments (Danesi et al., 2002).

As of January 19, 2011 in 30 countries 442 nuclear power plant units with an installed electric net capacity of about 375 GW are in operation and 65 plants with an installed capacity of 63 GW are in 16 countries under construction. Although there has been an increase in the decommissioning and the shutdown of nuclear reactors, nuclear energy production is still growing. Over 200 radionuclides are produced during the operation of a typical reactor; most of the radionuclides are relatively short-lived and decay to low levels within a few decades (Crowly, 1997). Neutron activation products and, to a less extent, volatile fission products such as iodine isotopes may enter the surrounding environment through atmospheric and aquatic releases from the nuclear power plants.The activity of radioactive material released worldwide in airborne and liquid effluents from reactors during routine operation for the period 2000-2002 is summarized in Table 4.

At present, the worldwide capacity to reprocess spent nuclear fuel is between one-third and one-half of its annual production rate (Hu et al., 2010). The main commercial reprocessing facilities are La Hague plant in France, which treats spent nuclear fuel from France, Japan, Germany, Belgium, Switzerland and Netherlands and thermal oxide reprocessing plant in Sellafield, United Kingdom, which reprocess fuel from United Kingdom, Japan, Germany and Switzerland. There are several others major reprocessing plants in the world, such as Mayak facility in Russia, Kalpakkam plant in India, reprocessing plant in Rokkasho, Japan, reprocessing plant in Lanzhou in

China and others. Extensive studies conducted in the area of Sellafield reprocessing plant have shown that discharged radionuclides accumulate on saltmarshes through a range of natural geochemical processes such as mixing and tidal currents (Assinder et al., 1985; Livens et al., 1994) and on tide-washed pastures bordering the Irish Sea coast of England and Wales (Sanchez et al., 1998).

Table 4. The radioactive material released worldwide in airborne and liquid effluents from reactors during routine operation

Radioactive material	Activity (GBq)		
	2000	**2001**	**2002**
Airborne effluents			
Noble gases	8.28×10^6	7.04×10^6	6.36×10^6
Tritium	3.42×10^6	2.91×10^6	3.14×10^6
Iodine-131	46.6	38.6	46.6
Carbon-14	4.42×10^4	4.18×10^4	5.26×10^4
Particulates	191	128	128
Liquid effluents			
Tritium	7.48×10^6	8.62×10^6	8.30×10^6
Other radionuclides	1.60×10^3	1.61×10^3	1.30×10^3

UNSCEAR, 2010.

Table 5. Estimates of principal radionuclides released in the Chernobyl accident

Radionuclides	Activity released (PBq)
Inert gases	
^{85}Kr, ^{133}Xe	6533
Volatile elements	
^{129m}Tc, ^{132}Te, ^{131}I, ^{133}I, ^{134}Cs, ^{136}Cs, ^{137}Cs	4228
Intermediately volatile	
^{89}Sr, ^{90}Sr, ^{103}Ru, ^{106}Ru, ^{140}Ba	606
Refractory	
^{95}Zr, ^{99}Mo, ^{141}Ce, ^{144}Ce, ^{239}Np, ^{238}Pu, ^{239}Pu, ^{240}Pu, ^{241}Pu, ^{242}Cm	693

UNSCEAR, 2010.

Nuclear wastes arise at various stages in the nuclear fuel cycle. They include low and intermediate-level wastes, mainly from reactor operation, high-level waste from fuel reprocessing, and spent fuel for direct disposal (UNSCEAR, 2010). Nuclear waste contains many radionuclides that, if released to the environment through repository failure, could potentially cause adverse effects in ecosystems (Nykyri et al., 2008).

Nuclear accidents can occur wherever radioactive materials are used, stored, or transported. In addition to nuclear power plants, hospitals, research laboratories, industries, nuclear powered vessels, nuclear icebreakers, satelites, lighthouses could be the site of a radiological accident. The 1986 accident at Chernobyl nuclear complex in Ukraine was the most severe ever to have occurred in the civilian nuclear power industry. It was estimated that 12 EBq of radioactivity was released in the Chernobyl accident (UNSCEAR, 2010). The principal radionuclides and their inventories are given in Table 5.

The latest serious accident occurred on March 12, 2011, in the Fukushima Daichi nuclear power plant in Japan, as a consequence of an earthquake in the Pacific Ocean followed by tsunami. Following the continuing air releases of radionuclides after the accident, several fission products were recorded in the air by a number of monitoring stations. In Europe, radioiodine (^{131}I) in air and rainwater as well as ^{134}Cs and ^{137}Cs were recorded in Thessaloniki, Greece (Manolopoulou et al., 2011). The elevated concentrations of several fission products (^{131}I, ^{132}I, ^{132}Te, ^{134}Cs and ^{137}Cs) were detected allong the Iberian Peninsula. The analysis of back-trajectories of air masses demonstrated that detected radionuclides are derived in Fukushima accident (Lozano et al., 2011). Traces of short- and long-lived fallout isotopes (^{131}I, ^{134}Cs and ^{137}Cs) were found in environmental samples collected in Northwest Germany (rain water, river sediment, soil, grass and cow milk) following the radioactivity releases after the nuclear accident in Fukushima (Pittauerová et al., 2011).

3. Geochemistry of Selected Radionuclides

The physical characteristics of soils are governed by many factors of which some include parent rock material, mode of formation of the minerals, the means and distance of transport, and the depositional environment. These characteristics together with the uranium and radium content of the parent rock, and recent physical and chemical events (i.e. chemical leaching, transport with water, and precipitation/adsorption) can affect the final distribution of radionuclides in the soil (Kemski et al., 1992).

Uranium, thorium and radium are often non-homogeneously distributed in rock material. The distribution depends primarily on the rock type (igneous, metamorphic, or sedimentary) and the extent of weathering. Uranium and thorium distributions differ due to their different chemical properties. According to Mason and Moore (1982) average crustal abundances of uranium and thorium are 1.8 mg kg^{-1} and 7.2 mg kg^{-1}, respectively.

In igneous rocks, uranium concentration increases with degree of differentiation, very low uranium concentrations occur in ultrabasic rocks (0.014 mg kg^{-1}) and higher in granites (2-15 mg kg^{-1}) and pegmatites (Rogers and Adams, 1969). Tieh et al. (1980) attributes uranium to three principal modes: background (uranium in major rock- forming minerals, such as quartz, feldspars, biotite and amphibolites, usually low concentrations levels), resistate (uranium in accessory minerals, such as zircon, apatite and sphene) and interstitial (uranium concentrated along grain boundaries and fractures in noncrystalline phases, formed during deuteric alteration and early weathering). Thorium in rocks is found at similar sites as uranium, as a trace constituent in solid solution in phosphate, oxide and silicate minerals.

In metamorphic rocks the uranium and thorium abundance and distribution is dependent on the source rock. The contents of both generally decrease with increasing degree of metamorphism (Rogers and Adams, 1969).

In sedimentary rocks uranium concentration increase with content of clay, phosphorus and organic matter. Sandstones, shales and limestones generally have low uranium contents, while phosphorites often contain more than 50 mg kg^{-1} of uranium. The uranium content of black shales can be as high as 1200 mg kg^{-1} (Rogers and Adams, 1969) and is often correlated to the content of organic material and abundance of colloidal size grades (Mason and Moore, 1982).

Radium in rocks is generally generated by the decay of uranium. Radium is not affected by long-term processes, due to its continuous production and its relatively short half-life of 1600 years. If no separation of radionuclides and their descendants takes place for 10000 years, ^{226}Ra geochemistry is governed entirely by ^{230}Th. However, in nature even short term equilibrium may not be reached because of the selective leaching of radium or thorium, or due to uranium enrichment (Molinari and Snodgrass, 1990).

Potassium tends to be concentrated in evolved crustal rocks such as granites and shales, where it is an essential constituent of several common-forming minerals (Van Schmus, 1995). Concentrations of uranium, thorium and potassium in major rock types are summarized in Table 6.

Table 6. Summary of concentrations of uranium, thorium and potassium in major rock types

Rock type	Uranium (mg kg^{-1})	Thorium (mg kg^{-1})	Potassium (%)
Igneous rocks			
Basalt	0.5-1	3-4	0.8
Mafic	0.5-0.9	1.6-2.7	0.3-1.1
Salic	3.9-4.7	16-20	4.5
Granite	3	17	>4
Sedimentary rocks			
Shale sandstones	3.7	12	2.7
Clean quartz	<1	<2	<1
Dirty quartz	2-3	3-6	2
Arkose	1-2	2	2-3
Beach sands	3	6	<1
Carbonate rocks	2	2	0.3
Continental upper crust	2.8	10.7	2.8

NCRP, 1987.

3.1. Uranium

Natural uranium consists primarily of three isotopes, ^{234}U (0.0057%), ^{235}U (0.71%) and ^{238}U (99.3%). Uranium-234 ($T_{1/2}$=2.5×10^5 years) decays through a series of radionuclides to the stable ^{206}Pb, simultaneously emitting alpha and beta particles and gamma rays. In the uranium decay chain ^{222}Rn ($T_{1/2}$=3.8 days) is one of the radionuclides of concern because it can easily enter the human lungs. Uranium-235 ($T_{1/2}$=7×10^8 years) and ^{238}U ($T_{1/2}$=4.5×10^9 years) decay to stable ^{207}Pb, and ^{206}Pb, respectively, emitting alpha and beta particles and gamma rays. Although uranium can be found as U^{4+}, U^{5+} and U^{6+}, U^{4+} and U^{6+} are the most likely uranium oxidation states and they are present in reducing and oxidizing conditions, respectively. Soil uranium taken up by plants is one of the primary pathways leading to human exposure. Uranium tends to concentrate in the bones and bone marrow, while when inhaled can provide a significant dose to the lungs. The biological half-life of uranium is 300 days for the bones and 100 days for general contamination of the body; however, the chemical toxicity of uranium is generally more significant than its radiotoxicity (Zhang et al., 2002).

It has been estimated that more than five percent of all known minerals contain uranium as an essential constituent (Mandarino, 1999). The number of discrete mineral phases identified is almost 200. These comprise U^{4+} oxides, silicates, phosphates and carbonates and a wide range of U^{6+} phases containing common and less abundant groundwater anions (Burns and Finch, 1999).

The solubility of uranium in soil is dependent on several factors such as: pH, redox potential, temperature, soil texture, organic and inorganic compounds, moisture and microbial activity (Rivas, 2005). Soluble forms can migrate with soil water, be uptaken by plants or aquatic organisms or volatilised (Igwe et al., 2005).

Uranium is a relatively mobile element in the surface soils owing to the stability of U^{6+} aqueous complexes. It may be precipitated by reduction to U^{4+} or in the form of uranyl minerals, principally phosphates, silicates, arsenates, vanadates and oxyhydroxides, several of which may occur simultaneously at the same locality (Burns and Finch, 1999). Adsorption/desorption behaviour of uranium has been a subject of numerous studies, including mechanistic studies of uranium uptake by oxides and hydroxides (Hsi and Langmuir, 1985; *Waite*et al.,1994), clay minerals and silica (Stamberg et al., 2003; Payne et al., 2004) and organic matter (Shanbag and Choppin, 1981). The bacterial reduction of soluble U^{6+} to insoluble U^{4+} is an important process for uranium fixation in soil (Anderson et al., 1989; Klimkhanner and Palmer, 1991). The investigations of potential of bacteria for bioaccumulation of uranium have been indicated accumulation of uranium in the form of sparingly soluble phoshate compounds (Selenska-Pobell et al., 2002).

The behavior of uranium in soils is complex. Uranium can be associated with numerous minerals and phases in soils, such as sulphates, phosphates, carbonates, oxides, hydroxides and clay minerals. The mobility of uranium in soil and its vertical transport (leaching) to groundwater depend on properties of the soil such as pH, oxidation–reduction potential, concentration of complexing anions, porosity of the soil, soil particle size and sorption properties, as well as the amount of water available (Allard et al., 1982; Bibler and Marson, 1992). Retention of uranium by the soil may be due to adsorption, chemisorption, ion exchange or a combination of mechanisms (Allard et al., 1982).

Uranium interacts with all components of the soil matrix such as clay minerals, aluminium and iron oxides, organic matter and micro-organisms. The bivalent uranyl ion (UO_2^{2+}) is sorbed to the negatively charged surfaces of clay minerals, sesquioxides and organic compounds. With increasing pH, more negatively charged binding sites are available on mineral surfaces due to the release of protons. On the other hand, with rising pH, there is a trend to increasing carbonate concentrations which is the most important complexing agent for uranium. Above pH 6, the fraction of uranium that is complexed by carbonates increases, which tends to enhance the uranium mobility in soil (Koch-Steindl and Pröhl, 2001). The organic substances of the soil solution bind uranium into complex compounds with different solubility and stability (Gu et al., 2005; Bednar et al., 2007). Sheppard et al. (1989) reported that background uranium concentrations in soils tend to increase with cation exchange capacity and/or organic matter content. Organic matter and clay minerals provide exchange sites and as such are expected to increase sorption of UO_2^{2+} and other positively charged uranium species. In soils with high phosphate content, U^{6+} may precipitate as uranyl phosphate (Duff and Amrhein, 1996).

3.2. Radium

Radium (^{226}Ra) is the decay product of ^{238}U, with a half-life of 1600 years. Radium exists in the environment typically as a divalent cation. Compared with uranium, radium has a higher mobility in soils and higher accessibility to the food chain (Zhang et al., 2002).

Radium has a distinctive chemistry compared with the actinides since it is an alkali earth element. It has been shown that radium sorbs easily onto natural soils and sediments (Benes et al., 1986) and onto almost all of their primary constituents, including ferric hydroxides, quartz, muscovite, albite, clay minerals, carbonates, aluminium oxides and organic matter (Benes et al., 1984; 1985; 1986a, b; Bassot et al., 2000; Deneux-Mustin, 2003). The degree of sorption depends on several variables owing to the multiple modes of interaction between radium and the soil mineral matrix. The most important ones are the ionic strength of the solution and the particle surface area. High concentrations of sulfate ion, low calcium concentrations, and low ionic forces of solutions restrain the migration of radium and favor its absorption in soils (Landa, 1980). The Ra^{2+} ion is weakly incline to complexing, though there are citric anionic complexes stable in an acid medium (Bagnall, 1957; Sedlet, 1966).An increase of mineral surface charge, the increased stability of inorganic radium complexes in high ionic strength solutions and the formation of strong organic complexes decrease radium sorption (IAEA, 2006). The biosorption experiments have shown that different biomass types, mainly fungi, accumulate large amounts of radium present in solution (Volesky, 1990).

Radium is metabolised similarly to calcium, and about 80% of the total body burden of radium is found in the skeleton, with the remainder being more or less uniformly distributed in the soft tissues (NCRP, 1987).

3.3. Thorium

Thorium has several isotopes whose atomic masses range from 212 to 236, but only one naturally occurring isotope, ^{232}Th ($T_{1/2}=1.4\times10^{10}$ years), which decays through emission of an alpha particle. Six alpha and four beta decay steps occur before ^{232}Th becomes stable ^{208}Pb. Thorium occurs in only one oxidation state (+4) in nature.

The main thorium containing minerals (thorite, thorianite, zircon and monazite) are resistant to weathering and do not dissolve readily in ground- or surface waters. Hydrous thorium oxide, the dominant form of Th^{4+} in aqueous environments, is known to precipitate in the natural environment and is likely to alter with time to a more crystalline solid that has a lower solubility. The solubility of hydrous Th^{4+} oxides decreases with increasing pH. It is rather immobile in the aqueous environment owing to the low solubility of thorianite but the high particle reactivity of thorium shows strong colloidal characteristics, hydroxide and silicate colloids allowing transport of thorium in soluble form. The movement of thorium is therefore controlled mainly by physical processes (USEPA, 1999).

Hydroxides of Th^{4+} are the dominant species in soil and aquatic systems, although carbonate complexes also form (Zhang et al., 2002). The global average for soil is given as 3.4 to 10.5 mg kg^{-1}. The mobility of thorium in soil, like uranium, is governed by the formation of the hydrated cation Th^{4+}, which is responsible for its solubility over a wide range of soil pH. Organic acids increase the solubility of thorium in soil, but its mobility may be limited due to the formation of slightly soluble precipitates (phosphates, oxides) and by adsorption on clay minerals and organic matter. Thorium has low mobility under all environmental conditions, mainly due to the high stability of the insoluble oxide ThO_2 and the strongly resistate nature of its carrier minerals such as monazite and zircon.

Thorium has no known biological function. It is chemotoxic, radiotoxic and a carcinogen. In the environment, thorium behaves similarly to the rare earth elements, especially cerium, substituting for calcium in bones and teeth.

Long-term exposure to thorium increases the chances of developing lung diseases and lung, pancreas and bone cancer (UNSCEAR, 2010).

3.4. Potassium

Of the naturally occurring potassium isotopes, only ^{40}K ($T_{1/2}$=1.3×10^9 years) is unstable. It decays by beta-particle emission to ^{40}Ca (89%) and by electron capture to ^{40}Ar (11%) and produces 1.46 MeV gamma rays after electron-capture decay. Potassium-40 is present at 0.0117% by mass in natural potassium. Because of its relative abundance and its energetic beta-particle emission (1.3 MeV), ^{40}K is predominant radioactive component in common foods and human tissues.

Potassium is widely distributed in the Earth's crust with concentrations ranging from about 0.1% for limestones to more than 4% for some granites. For crustal rock, the mean ^{40}K activity is 620 Bq kg^{-1}, with basalt averaging about half this amount (Klement, 1982).

Potassium, and hence ^{40}K, is found in all biological systems. Next to radon and its daughter products, which primarily irradiate the respiratory tract, ^{40}K is predominant source of internal radiation dose from naturally occurring radioactive materials (UNSCEAR, 2010).

3.5. Caesium

Caesium fission products include three main isotopes at significant concentrations: ^{134}Cs ($T_{1/2}$=2.05 years), ^{135}Cs ($T_{1/2}$=3×10^6 years) and ^{137}Cs ($T_{1/2}$=30.2 years). The solubility of most caesium compounds is very high: precipitation reactions are not expected to control caesium's geochemical behaviour. Most soils sorb caesium rather strongly and often selectively. Some micaceous minerals, such as illite and vermiculite, tend to take up caesium in their intrastructural layers (Cremers et al., 1988). Thus, the high retention of caesium in soil is determined by two different processes: structural fixation and reversible selective sorption (Konoplev et al., 2002). Reversible selective sorption occurs on frayed edge sites, located at the edges of micaceous clay particles (Cremers et al., 1988). The interlayer distance in certain micaceous minerals excludes diffusion of some major ions while permitting the Cs^+ ion to fit perfectly between the layers. The low hydration energy of Cs^+ ion is considered to be the major factor in its fixation (Sawhney, 1972).

Caesium is strongly sorbed in soil by ion exchange, some of which is irreversible, or fixed, with fixation being influenced by clay mineralogy (Livens and Rimmer, 1988). A number of models relating the availability of radiocaesium to soil properties have been proposed including: increasing soil-plant transfer with increasing soil organic matter (Frissel et al., 1990; Wright et al., 2003), decreasing soil – plant transfer with increasing soil solution K^+ (Sanchez et al., 1999), and a mechanistic approach using soil clay and organic matter contents, exchangeable K^+ status, pH and NH_4^+ concentration (Absalom et al., 2000).

Because of its long half-life and chemical-biochemical similarity with potassium, ^{137}Cs is one of the most important radionuclides released into the environment, especially after the Chernobyl nuclear power plant accident.

4. Factors Affecting Radionuclide Concentrations in Soil

The important interactions of any radionuclide in soil solution, which can influence its mobility in soils and eventual plant root uptake, include: charge interactions, complexation and precipitation reactions with other chemical species, oxidation-reduction transformation, and specific interactions with soil components including soil biota (Beresford, 2006).

Variations of background concentrations of radionuclides in soil depends on the soil type, moisture content, inhomogeneity of its permeability, formation, transport processes and geomorphology associated with meteorological conditions (El-Arabi, 2005). Soil formation and chemical and biochemical interactions also influence the distribution patterns of uranium, thorium and their decay products (Karakelle et al., 2002). Radioactivity in soil results from the rock which it is derived, including porosity and water and organic matter, and increase by sorption and precipitation of radionuclides from incoming water.

The mobility and bioavailability of radioactive elements in the environment is also dependent upon the element's physico-chemical form. Therefore, it is essential to determine what chemical forms are present, i.e. to identify speciation of radionuclides of interest. In soils, speciation is generally related to the physico-chemical forms of the elements of concern, but also to the properties of the soil matrix: simple and complex ions in interstitial solution, exchangeable ions, associated with organic soil fractions, occluded or coprecipitated with metal oxides, carbonates or phosphates and other secondary minerals, and ions in the crystal lattices of primary minerals (Zhu and Alva, 1993).

The cationic species UO_2^{2+}, Th^{4+} and $Th(OH)_2^{2+}$ are the dominant forms of these elements in soils and are strongly sorbed on soil solids, including organic matter (Sheppard and Evenden, 1988). However, organic complexes and colloids can increase their mobility in mineral soils. Mortvedt (1994) noted that the plant uptake of thorium, uranium and radium was influenced by soil pH, and that this may be related to indirect effects of competition from calcium and magnesium, particularly for radium, which belongs to the alkaline earth series. It has been shown that clay content is the most important determinant of background levels of ^{40}K, ^{226}Ra, and ^{232}Th in soils (Van den Bygaart et al., 1999). There is evidence of increased association of the three radionuclides with the clay fraction compared with the sand fraction of soils contaminated by uranium mine tailings. Soil organic matter appears to form strong complexes with thorium and to increase its mobility in soil (Sheppard, 1980; Mortvedt, 1994). Selenium, available potassium, carbonate, phosphorus and nitrates, have been observed to influence plant uptake of uranium (Van Netten and Morley, 1983).

The leading mechanism of radiocaesium fixation in soils is its exchangeable and non exchangeable sorption and fixation (Korobova et al., 2007). It has been shown that ^{137}Cs sorption by soils is controlled by ion-exchange mechanisms. The compensation of the negative charge of clay particles by positively charged caesium ions depends on the structure of the particular minerals, their selectivity with respect to caesium ions, the size of the mineral particles and the degree of their aggregation, the presence of competitive ions, the pH and ionic strength of the soil solution, and the caesium concentration in it. The vertical migration of radiocaesium in soils is related to the rate of infiltration of water together with colloidal and clay particles carrying the radionuclides. Elejalde et al. (1996) have shown that ^{137}Cs activity concentration in soil is significantly related to formation of a complex with organic matter, substitution for K^+, H^+ or NH_4^+ and a competitive effect of ^{40}K for retention by soil with a small contribution due to soil texture. Bihari and Dezső (2008) noted that ^{137}Cs association is concentrated in the finest soil fractions but adsorbed almost entirely to clay minerals. Baeza et al. (1995) reported significant correlation between ^{137}Cs activity concentrations in soil and organic matter content. The lateral and vertical distribution of ^{137}Cs is also influenced by soil management such as tillage, soil erosion and soil organic matter cycling.

Soil biota play an important role in the movement of radionuclides through a soil system. Microbial diversity within soil is integral to soil ecosystem function (Torsvik and Øvreås, 2007). The complex and dynamic nature of natural microbial communities has a profound effect on the soil medium and therefore to the geochemical behaviour of radionuclides in soil (Fomina et al., 2007). Prokaryotes can be considered especially important since they represent the largest phylogenetic diversity of any grouping on Earth and are involved in all biogeochemical cycling (Torsvik and Øvreås, 2007). Fungi impact upon the structural dynamics of soil and any transport process that are affected by soil structure are also likely to be affected by the the effect of fungi on soil architecture (Ritz, 2006). Both bacteria and fungi immobilise and complex radionuclides by mechanisms such as volatilization, extra-cellular complexing, intracellular accumulation and cell surface binding (Ragnarsdottir and Charlet, 2000). Fungi accumulate radionuclides and are also able to form mycorrhizal links to vascular plants and thus enhance radionuclide uptake by vascular plant hosts (Haas et al., 1998).

The behaviour of radionuclides in soil controls their migration in soil, the possible transport to groundwater, and the long-term radionuclide concentration in plants and thus the exposure of soil organisms. The entry of radionuclides into human food chains is controlled by their uptake by plant roots.

5. Distribution of Radionuclides in Surface Soils from Serbia

Investigation of soil radioactivity receives particular attention worldwide. These results serve as reference information to assess any changes in background level due to any artificial influences on the environment. Baseline data of this type are of importance in making estimations of population exposures. In recent years, an extensive survey of soil radioactivity has been conducted in Serbia. The activity concentrations of ^{226}Ra, ^{232}Th, ^{40}K and ^{137}Cs in surface soils determined by gamma-ray spectrometry, according to procedure described in detail in previous papers (Dragović and Onjia, 2005; Dragović et al., 2005; Dragović et al., 2006b; Dragović and Onjia, 2007), are presented in Table 7.

Table 7. Descriptive statistics of activity concentrations of ^{226}Ra, ^{232}Th, ^{40}K and ^{137}Cs in surface soils from Serbia

Parameter	Activity concentration (Bq kg^{-1})			
	^{226}Ra	^{232}Th	^{40}K	^{137}Cs
Mean	33	36	508	41
Median	32	37	508	29
Mode	36	40	610	14
St. deviation	12	12	160	36
Minimum	8	9.5	45	2
Maximum	88	84	900	160

The spatial distribution of the ^{226}Ra, ^{232}Th, ^{40}K and ^{137}Cs activity concentrations in surface soils from Serbia obtained by ArcInfo and ArcGIS from ESRI (ESRI, 2009) is presented in figures 1-4. The natural radionuclides showed a range of concentrations, as a consequence of the variety of lithological components in the investigated area (Dragović et al., 2006a, b). Geologically, the territory of Serbia and Montenegro includes a great number of rock complexes (magmatic, sedimentary and metamorphic rocks) which are markedly different with respect to age, genesis, mineral content and petrochemical and geochemical characteristics. Outstanding differences in natural radioactivity of soils can exist in relation to their geological origin (Dimitrijević, 1997). The territory of Serbia is divided into five regional geotectonic units: Pannonian Basin, Interior Dinarides, Carpatho-Balkanides, Vardar Zone and Serbian-Macedonian Mass, which are significantly different according to types of water-bearing media (Dimitrijević, 1997; Dimitrijević, 2002). The Pannonian Basin, the northern unit which occupies over 20% of the teritorry of Serbia, is composed of Tertiary and Quaternary sediments. The Dinarides, extended in the Western parts of Serbia, consist dominantly of fractured and karstified Mesozoic limestones. The Carpatho-Balkanides in the eastern part of Serbia are characterized by masses of Jurassic and Cretaceous limestones, with numerous uranium mineralizations. The Vardar Zone is composed of several blocks of diverse composition, geological history and provenience, and it includes characteristic oceanic elements. This zone, located to the west of the Serbian-Macedonian Mass, consists of a composite assemblage of continental and oceanic-derived units, including also Triassic and Jurassic ophiolites. The Vardar Zone is the most complex zone im the Balkan Peninsula, as a belt of multiple opening and closing of oceanic areas.

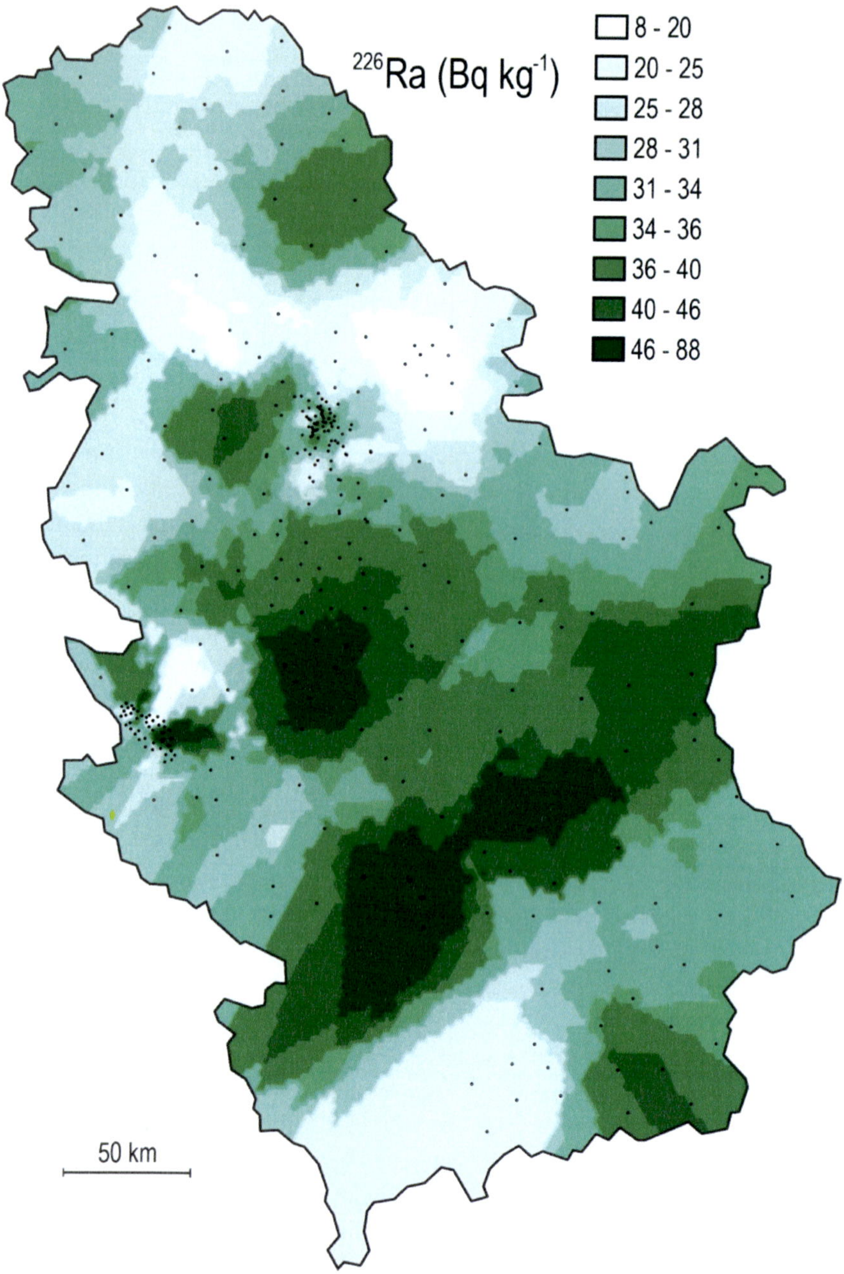

Figure 1. Spatial distribution of the ^{226}Ra activity concentrations in surface soils from Serbia

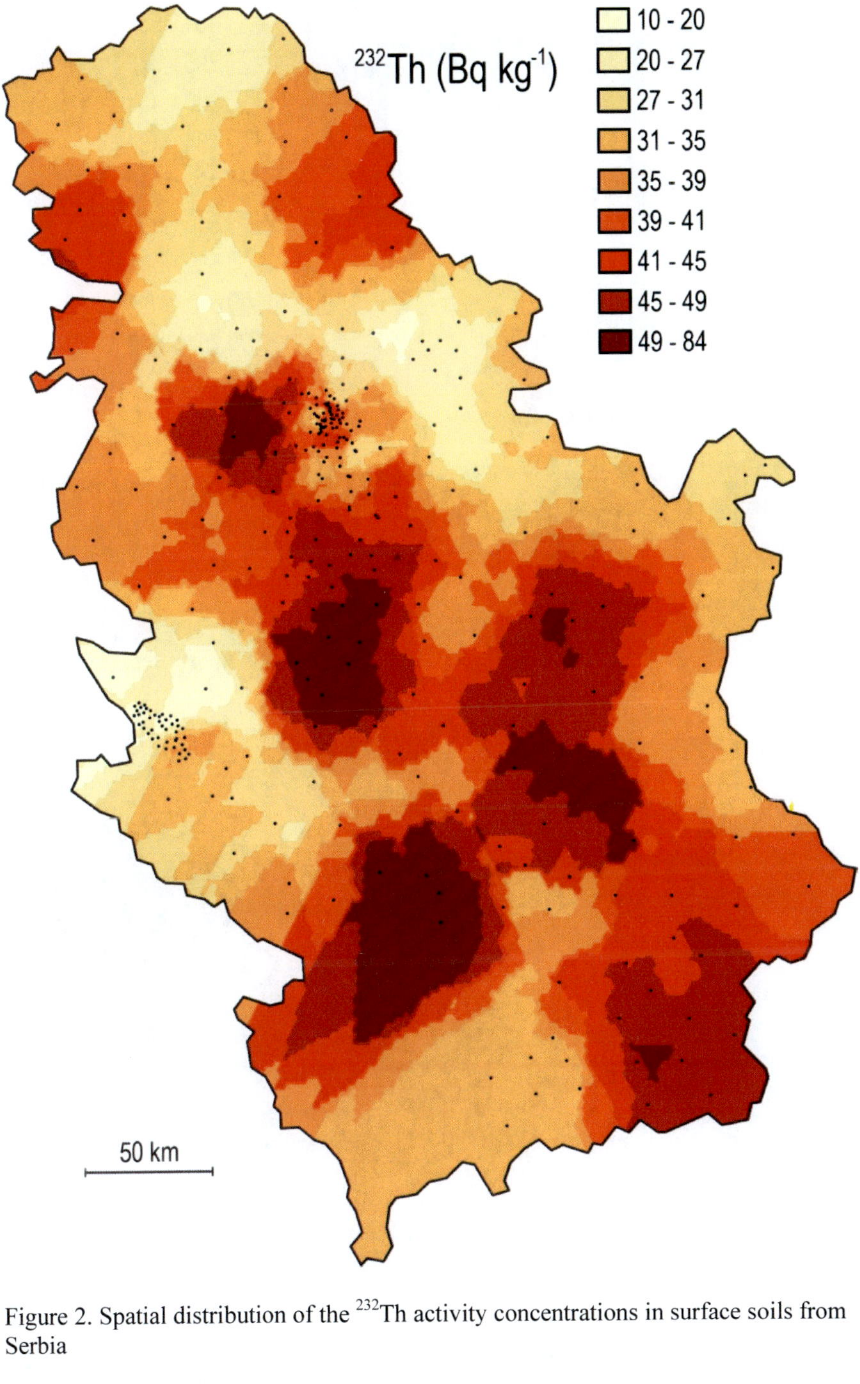

Figure 2. Spatial distribution of the ^{232}Th activity concentrations in surface soils from Serbia

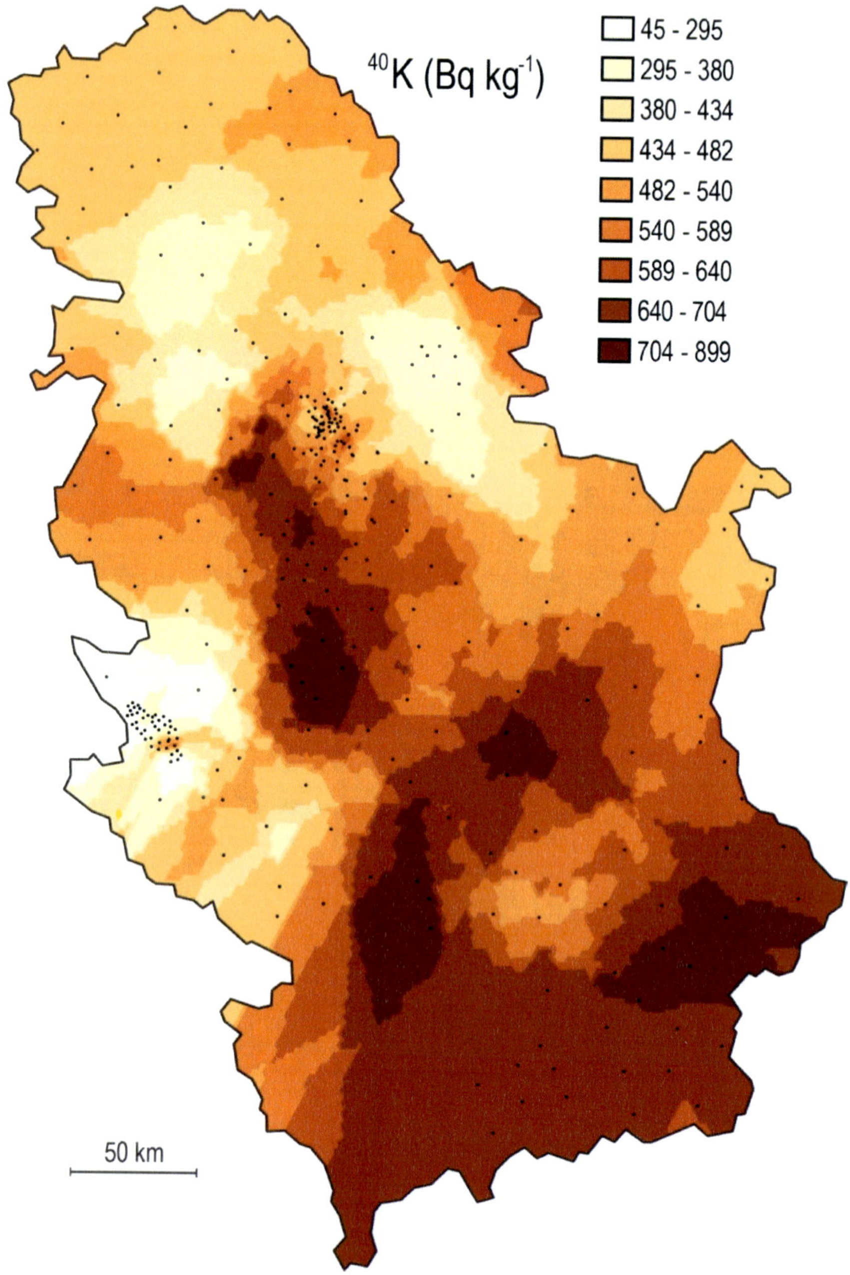

Figure 3.Spatial distribution of the ^{40}K activity concentrations in surface soils from Serbia.

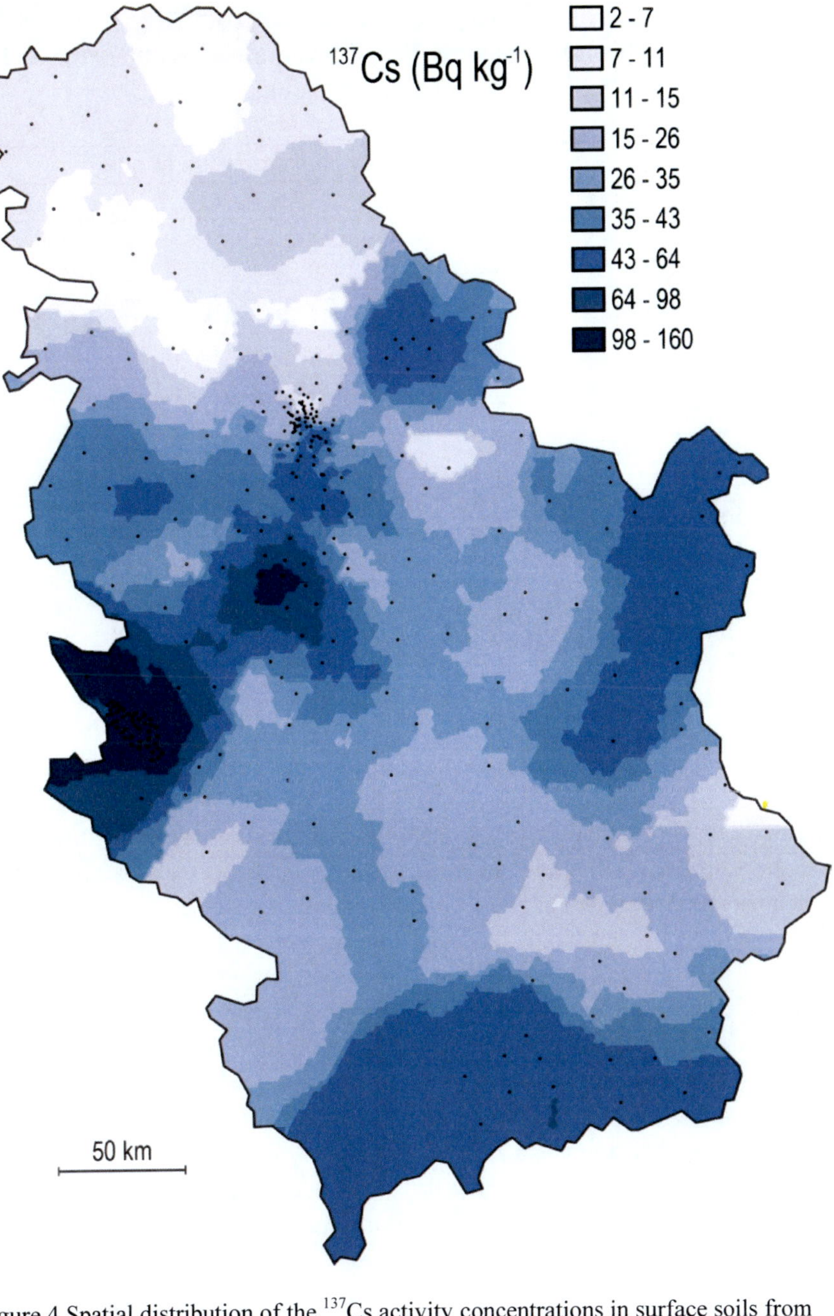

Figure 4.Spatial distribution of the ^{137}Cs activity concentrations in surface soils from Serbia.

Table 8. The comparison of activity concentrations of ^{226}Ra, ^{232}Th and ^{40}K in surface soils from Serbia with the values obtained in other studies in East and South Europe

Country	Activity concentration ($Bq\ kg^{-1}$)		
	^{226}Ra	^{232}Th	^{40}K
Albania	-	18	348
Bulgaria	45	30	400
Croatia	43	37	423
Cyprus	17	-	140
Czech Republic	44	41	613
FYR Macedonia	23	25	456
Greece	29	28	383
Hungary	33	28	370
Montenegro	29	24	246
Poland	25	25	410
Romania	32	38	490
Russian Federation	27	30	520
Slovakia	32	38	520
Slovenia	41	35	374
Serbia(this chapter)	33	38	547

UNSCEAR, 2010 and references cited therein.

The Serbian Macedonian Mass, located between the Vardar Zone in the west, and the Carpatho-Balkanides in the east, is composed of crystalline schists. The mean radionuclides activity concentrations in surface soils from Serbia were found to be similar to those for surface soils worldwide (UNSCEAR, 2010). The high activity levels of ^{226}Ra were observed in granitoid and volcanic complexes of Serbia where uranium occurrs in its own minerals formed during magma crystallisation, in accessory minerals of granitoid complexes, and in minerals formed as a consequence of hydrothermal activities. The volcanic complexes which spread over a significant part of the territory of Serbia are characterised by anomalous uranium concentrations. The majority of occurrences and deposits of uranium are associated with clastic sedimentary rocks, e.g. fine-grained sandstones with high organic matter content, particularly in basins close to granitoid complexes. The comparison of the activity concentrations of ^{226}Ra, ^{232}Th and ^{40}K in surface soils from Serbia with those obtained worldwide is presented in Table 8. The spatial distribution of ^{137}Cs on the territory of Serbia was influenced by geographical factors, such as altitude, precipitation rate and variations in stratospheric circulation. The highest ^{137}Cs activity concentrations in soils from Serbia were observed in highlands which receive the highest precipitation.

The correlation between natural radionuclides in soil can be used to deduce some conclusions about coexistence of the radionuclides and may provide an indication whether their relative depletion or enrichment had occured. Chemically, ^{226}Ra, ^{232}Th and ^{40}K behave very differently (Hansen and Huntington, 1969; Sadusky et al., 1987). However, any correlations between radionuclide concentrations allow some conclusions about the likely location of the radionuclide to be drawn, in particular whether the radionuclides are internally bound within silicate minerals, or associated with surface coatings. The correlations between ^{226}Ra and ^{232}Th, ^{226}Ra and ^{40}K, and ^{232}Th and ^{40}K in surface soils from Serbia are presented in Figure 5. In all three cases poor correlation was observed with correlation coefficients of 0.654, 0.388 and 0.662, respectively. The poor correlation coefficients indicate that the individual results for any of the radionuclide concentrations in each pair are not a good predictor of individual values for the other (Prakash et al., 2007; Shiva Prasad et al., 2008; Harb, 2008). The poor correlation between radionuclides shows their association with different chemical and geological forms. The soil particle size fractions also contribute to the correlation between radionuclides, with macro-cracks, fractures and mobility of the radionuclides in geological medium (Ramola et al., 2011).

The absorbed outdoor gamma dose rate (nGy h^{-1}) in the air at 1 m the above ground level was calculated from the measured activity concentrations of ^{226}Ra, ^{232}Th and ^{40}K in soil assuming that other radionuclides, such as ^{137}Cs,^{90}Sr and the ^{235}U series can be neglected as they contribute very little to the total dose from environmental background (Kocher and Sjoreen, 1985; Jacob et al., 1986; Leung et al., 1990). The calculation of gamma dose rates due to ^{226}Ra, ^{232}Th, ^{40}K and ^{137}Cs were performed applying the following conversion coefficients: 0.462, 0.604, 0.042, 0.066, respectively (Tahir et al., 2006; UNSCEAR, 2010). The mean outdoor gamma dose rate due to natural radionuclides in soils from Serbia (Table 9) was slightly higher than worldwide average of 54 nGy h^{-1} (UNSCEAR, 2010). Tha gamma dose rate due to ^{137}Cs contributed only 4.4% to the total gamma dose rate.

The estimation of population exposure and establishment of baseline data for further environmental assessments are of special importance for cities with large populations. With a growing *population of 2 million* people, Belgrade is the capital and largest city of Serbia, which is expected to develop from a city into a conurbation. The basic descriptive statistics of activity concentrations of ^{226}Ra, ^{232}Th, ^{40}K and ^{137}Cs in surface soils from Belgrade is given in Table 10.

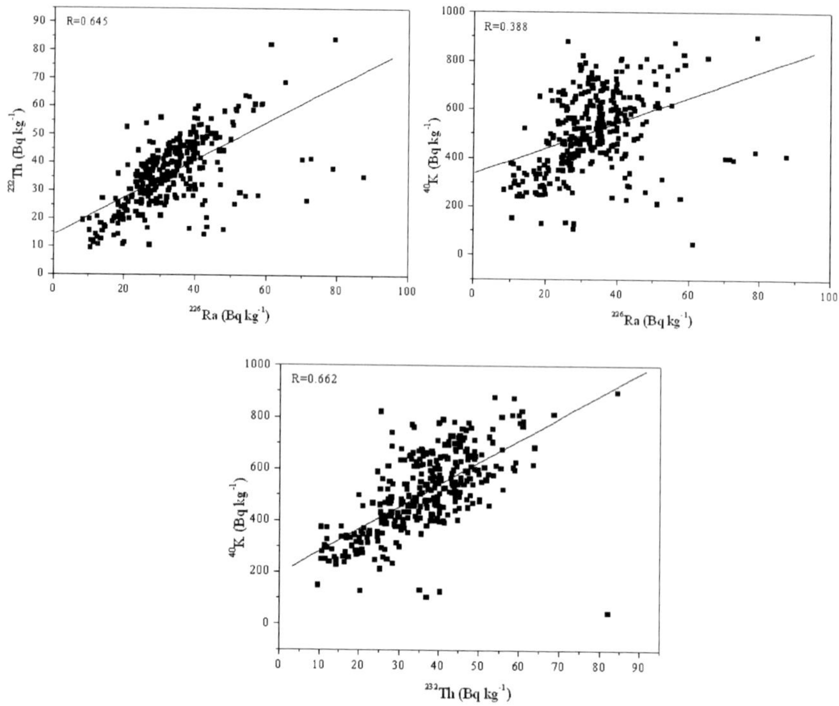

Figure 5.The correlations between activity concentrations of natural radionuclides in soils from Serbia.

Table 9. The absorbed outdoor gamma dose rates in air at 1 m above ground level due to ^{226}Ra, ^{232}Th, ^{40}K and ^{137}Cs in surface soils in Serbia

Parameter	Gamma dose rate (nGy h^{-1})				
	^{226}Ra	^{232}Th	^{40}K	^{137}Cs	Total
Mean	15	22	21	2.7	61
Median	15	22	21	1.9	61
Mode	13	24	23	0.3	51
St. deviation	6	8	7	2	16
Minimum	3.8	5.8	1.9	0.1	22
Maximum	40	51	38	11	127

Table 10. Descriptive statistics of activity concentrations of ^{226}Ra, ^{232}Th, ^{40}K and ^{137}Cs in surface soils from Belgrade

Parameter	Activity concentration (Bq kg^{-1})			
	^{226}Ra	^{232}Th	^{40}K	^{137}Cs
Mean	33	39	506	24
Median	35	39	510	16
Mode	35	39	527	12
St. deviation	7.8	9.0	95	19
Minimum	12	11	285	2.8
Maximum	55	64	756	87

The natural radionuclides activity concentrations differed among sampling locations as a consequence of the variety of geological structures in the investigated area caused by complex geotectonic processes which occured in several phases (Janković Mandić and Dragović, 2010; Janković Mandić et al., 2010). The spatial distribution of the ^{226}Ra, ^{232}Th, ^{40}K and ^{137}Cs activity concentrations in surface soils from Belgrade is presented in figures 6-9. Most sampling locations were situated on terrains of the southern rim of the Panonian area and the northern parts of the Sumadian mesozoic joist. There are two structural floors in the Belgrade area tectonics. The first is composed of jurassic and cretaceous sediments and magmatic products. Parts of the mesozoic joist on the west are composed of flysh sediments of the upper cretaceous series, which are covered by neogen sediments with horizontal stratification important for the development of pedological cover. The second structural floor is composed of neogen sediments. More than 70% of the territory of Belgrade is covered by quaternary deposits of alluvial lake, alluvial marsh, alluvial, delluvial, delluvial–prolluvial, prolluvial and eolic sediments. Near the Sava and Danube rivers, alluvial sediments are deposited, while loess is predominant in the western parts of Belgrade (Komatina and Komatina, 1999). Such a geological composition, together with the expressed structural formations and different terrain elevation, leads to differences in radionuclide concentrations in soils.

The absorbed outdoor gamma dose rates due to ^{226}Ra, ^{232}Th, ^{40}K and ^{137}Cs in surface soils from Belgrade (Table 11) were found to be similar to those obtained for whole Serbian territory. The contribution of ^{137}Cs to the total gamma dose rate (2.6%) lower than that for whole Serbian territory could be attributed to the lower altitudes of sampling sites in Belgrade.

Table 11. The absorbed outdoor gamma dose rates in air at 1 m above ground level due to ^{226}Ra, ^{232}Th, ^{40}K and ^{137}Cs in surface soils from Belgrade

Parameter	Gamma dose rate (nGy h^{-1})				
	^{226}Ra	^{232}Th	^{40}K	^{137}Cs	Total
Mean	15	23	21	1.6	61
Median	16	24	21	1.1	62
Mode	16	24	23	0.8	25
St. deviation	3.6	5.4	3.9	1.2	12
Minimum	5.3	6.8	12	0.2	25
Maximum	25	38	32	5.7	92

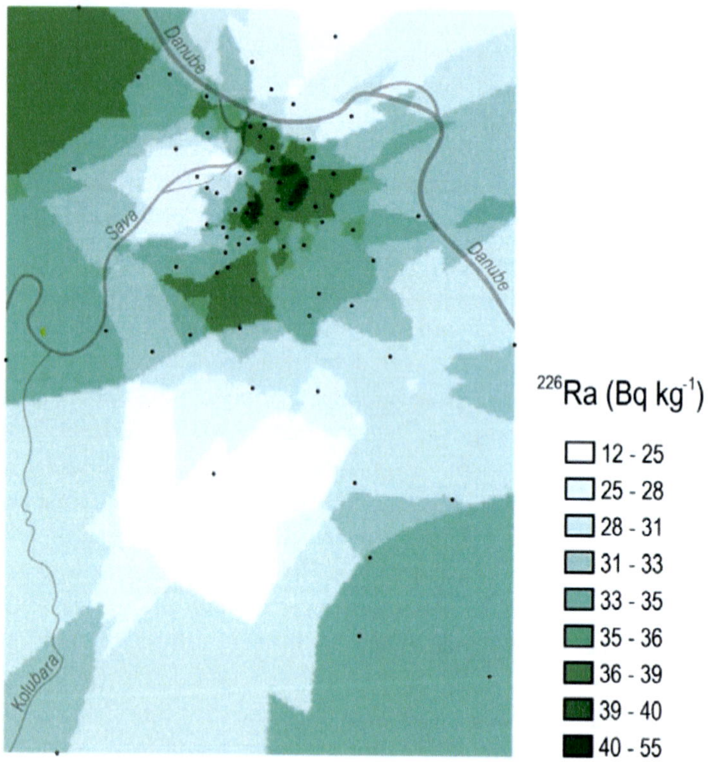

Figure 6. Spatial distribution of the ^{226}Ra activity concentrations in surface soils from Belgrade

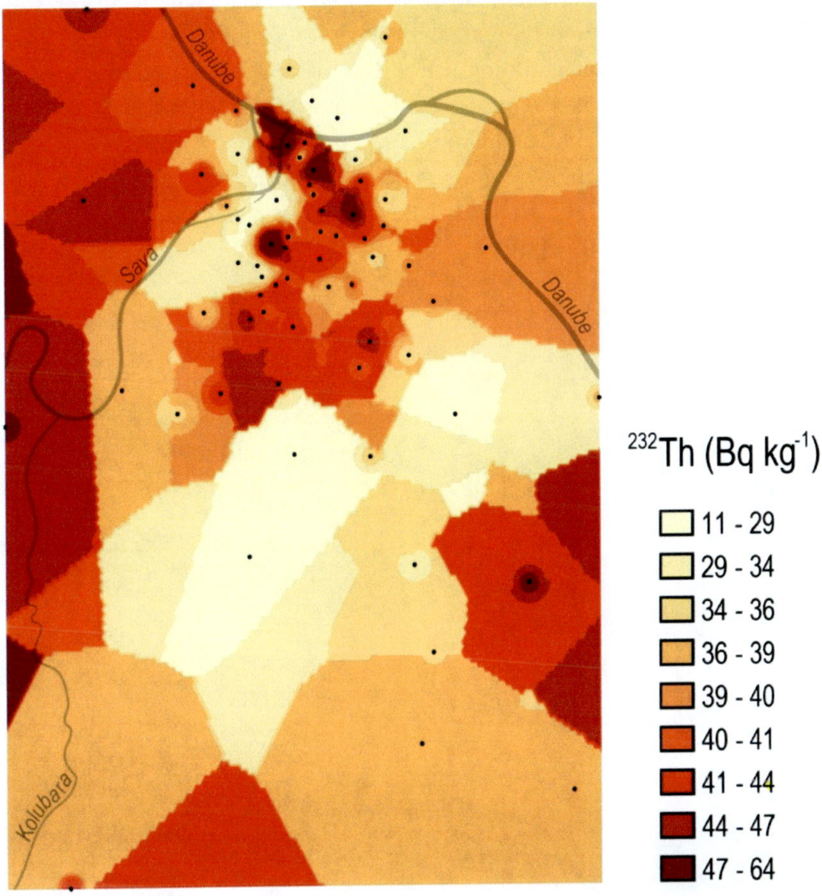

Figure 7. Spatial distribution of the ^{232}Th activity concentrations in surface soils from Belgrade.

The relatively high ^{137}Cs activity concentrations have been observed for highlands in Serbia. In soils collected twenty years after Chernobyl accident at Zlatibor Mountain the ^{137}Cs activity concentrations varied in range from 73 to 160 Bq kg^{-1}, with a mean of 108 Bq kg^{-1}. Their spatial distribution is presented in Figure 10. The high ^{137}Cs activity concentrations in analysed soils were influenced by high altitudes of sampling sites. Forest ecosystems on higher altitudes are predisposed to receive more fallout and therefore higher concentrations of radionuclides (Howard et al., 1991). This can be explained by the fact that annual means of precipitation under European conditions mostly rise with altitude. Semi-natural ecosystems present almost closed biogeocenoses, with efficient cycling of nutrients, where the level of radioactive contamination remains high for a long time (Baldini et al., 1989). The high activity concentration of ^{137}Cs in soil even two decades after Chernobyl accident have been reported for other highlands in Serbia (Dragović and Onjia, 2006; Dragović and Onjia, 2007). High ^{137}Cs activity concentration in soils of two mountainous regions in neighbouring Bulgaria have been noted, i.e. 288-827 Bq kg^{-1} for Rila Mountain and 86-2543 Bq kg^{-1} for Central Balkan (Zhiyanski et al., 2008).

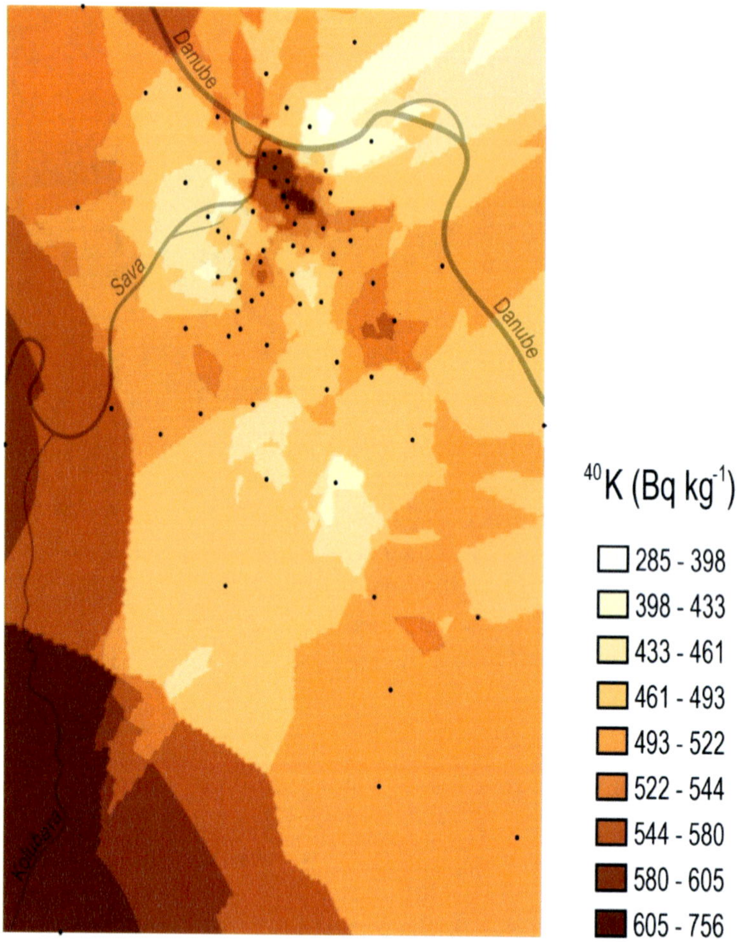

Figure 8. Spatial distribution of the ^{40}K activity concentrations in surface soils from Belgrade

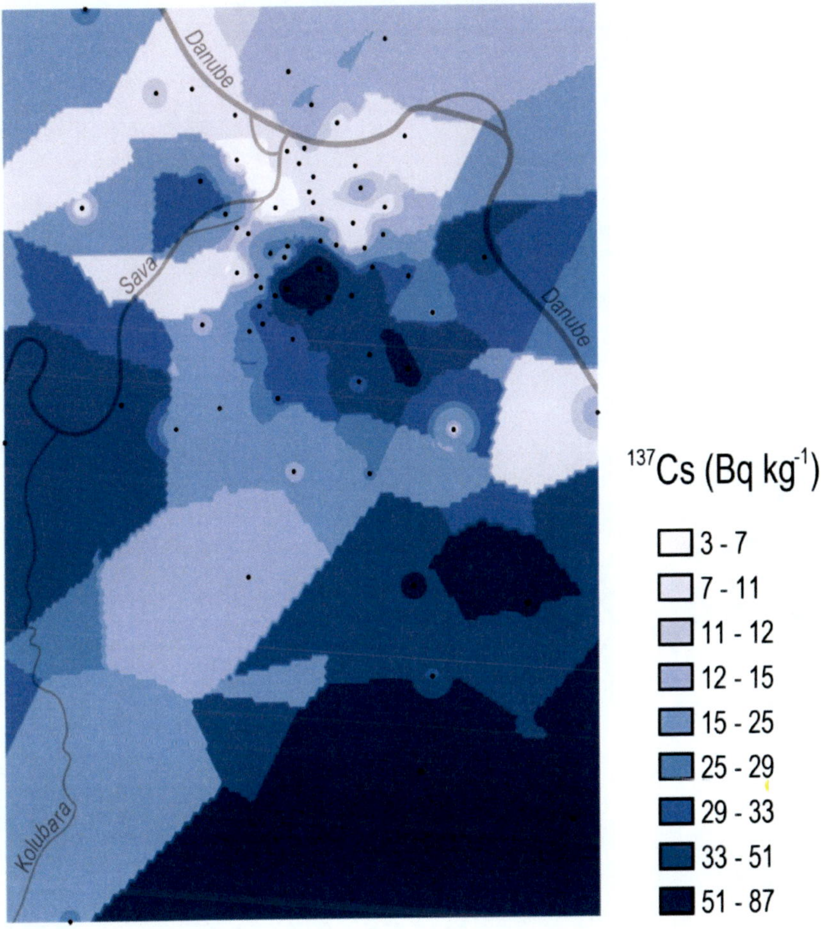

Figure 9. Spatial distribution of the ^{137}Cs activity concentrations in surface soils from Belgrade.

Radioecological sensitivity to radiocaesium in semi-natural ecosystems is generally higher than in agricultural ecosystems, sometimes by a few orders of magnitude (Howard et al., 2001; UNSCEAR, 2010). This difference is caused by a number of factors such as the multilayered vertical structure of soils in semi-natural ecosystems characterised mainly by a clay-poor mineral layer; the lack of competition between caesium and potassium which results in higher transfer rates for radiocaesium in nutrient-poor ecosystems, the presence of specific food-chain pathways that lead to high radionuclide activity concentrations in products from semi-natural ecosystems.

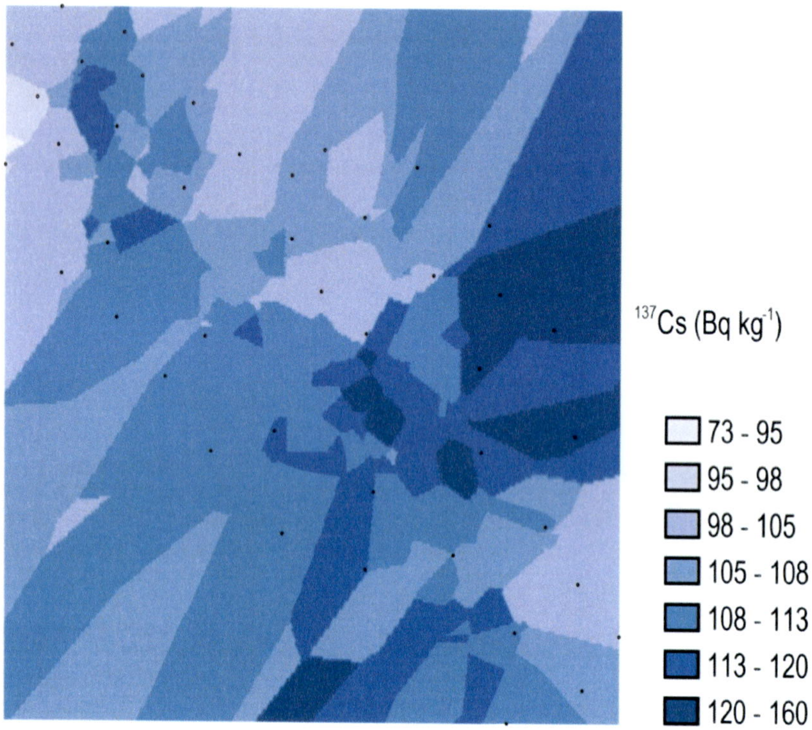

Figure 10.Spatial distribution of ^{137}Cs activity concentrations in surface soils from Zlatibor Mt.

ACKNOWLEDGMENTS

This work was supported by Ministry of Education and Science of the Republic of Serbia (Project No. III43009). The authors are grateful to Dr Heleny Florou, NCSR Demokritos, Athens, who reviewed this chapter, for her careful reading of the text and helpful remarks.

REFERENCES

Absalom, J.P.; Young, S.D.; Crout, N.M.J.; Sanchez, A.; Wright, S.M.; Smolders, E.; Nisbet, A.F.; Gillett, A.G. *J. Environ. Radioactiv.* 2000, 52, 31-43.

Al-Jundi, *J. Radiat. Meas*. 2002, 35, 23-28.

Allard, B.; Olofsson, U.; Torstenfelt, B.; Kipasti, H.; Andersson, K. *Mater. Res. Soc. Symp. Proc*. 1982, 11, 775–782.

Anderson, R.F.; Lehuray, A.P.; Fleisher, M.Q.; Murray, J.W. *Geochim.Cosmochim. Ac*. 1989, 53, 2205-2213.

Assinder, D.J.; Kelly M.; Aston, S.R. *J. Environ. Radioactiv*. 1985,2, 1–22.

Baeza, A.; Del Río, M.; Jiménez, A.; Miró, C.; Paniagua, J. *Radiochim. Acta* 1995, 68, 135-140.

Bagnall, K.W. Chemistry of the Rare Radioelements; Academic Press: New York, 1957.

Baldini, E.; Bettoli, M.G.; Tubertini, O. *Radiochim. Acta* 1989, 46, 143–144.

Bassot, S.; Stammose, D.; Mallet, C.; Ferreux, J.-M.; Lefebvre, C. In Proc. 10th Int. Conf. Hiroshima 2000; International Radiation Protection Association (IRPA): Paris, 2000.

Bednar, A.J.; Medina, V.F.; Ulmer-Scholle, D.S.; Frey, B.A.; Johnson, B.L.; Brostoff, W.N.; Larson, S.L. *Chemosphere* 2007, 70, 237-247.

Benes, P.; Borovec, Z.; Strejc, P. *J. Radioanal.Nucl*. Ch. 1984, 82, 275-285.

Benes, P.; Borovec, Z.; Strejc, P. *J. Radioanal.Nucl. Ch*. 1985, 89, 339-351.

Benes, P.; Borovec, Z.; Strejc, P. *J. Radioanal.Nucl. Ch*. 1986a, 98, 91-103.

Benes, P.; Borovec, Z.; Strejc, P. *J. Radioanal.Nucl. Ch*. 1986b, 99, 407-422.

Beresford, N.A. Soil Use Manage. 2006, 21, 468-474.

Bibler, J.P.; Marson, D.B. Behavior of mercury, lead, cesium, and uranyl ions on four SRS soils (U); Technical Report WSR-RP-92-326; Westinghouse Savannah River Company: Aiken, SC, 1992.

Bihari, Á.; Dezső, Z. *J. Environ. Radioactiv*. 2008, 99, 1083-1089.

Blanco, P.; Vera Tomé, F.; Lozano, J.C. *J. Environ. Radioactiv*.2005, 79, 315-330.

Borio, R.; Campos Venuti, G.; Risica, S.; Simula, S. *Sci. Total Environ*. 1985, 45, 55-62.

Burns, P.C.; Finch, R.; Eds.; Uranium – Mineralogy, Geochemistry and the Environment, Reviews in Mineralogy 38; Mineralogical Society of America: Chantilly, VA, 1999.

Carvalho, F.P.; Madruga, M.J.; Reis, M.C.; Alves, J.G.; Oliveira, J.M.; Gouveia, J.; Silva, L. *J. Environ. Radioactiv*. 2007, 96, 39-46.

Carvalho, F.P.; Oliveira, J.M.; Malta, M. J. *Radioanal.Nucl. Ch*. 2009, 281, 479-484.

Choppin, G.R. Radiochim. Acta 2003, 91, 645-649.

Clarke, L.B. Applications for Coal-use Residues; IEA CR/50; IEA Coal Research: London, 1992.

Clarke, L.B. Management of FGD Residues; IEA CR/75; IEA Coal Research: London, 1993.
*Cremers*A.; Elsen A.; De Preter P.; Maes A. *Nature 1988, 335,247-249.*
Crowly, K.D. Phys. Today 1997, 50, 32-39.
Danesi, P.R., Moreno, J., Makarewicz, M., Radecki, Z.*J. Radioanal. Nucl. Ch.* 2002, 253, 53-65.
Deneux-Mustin, S. Mobilité et transfert racinaire des éléments en traces, influence des micro-organismes du sol; Editions Tec et Doc: Paris, 2003.
Dimitrijević, M.D. Geology of Yugoslavia; Geoinsitute-Barex: Belgrade, 1997.
Dimitrijević, M.D.; Ed.; Geological Atlas of Serbia; The Serbian Ministry of Natural Resources and Environmental Protection: Belgrade, 2002; pp 187.
Dragović, S.; Janković, Lj., Onjia, A.; Bačić, G. *Radiat. Meas.* 2006a, 41, 611-616.
Dragović, S.; Janković, Lj.; Onjia, A. *Radiat. Prot. Dosim.* 2006b, 121, 297-302.
Dragović, S.; Onjia, A. *Appl. Radiat. Isotopes* 2005, 63, 363-366.
Dragović, S.; Onjia, A. *Appl. Radiat. Isotopes* 2007, 65, 218-224.
Dragović, S.; Onjia, A. *J. Environ. Radioactiv.* 2006, 89, 150-158.
Dragović, S.; Onjia, A. Russ.*J. Phys. Chem.* A 2007, 81, 1477-1481.
Dragović, S.; Onjia, A.; Bačić, G. *Nucl.Instrum.Meth.* A 2006c, 564, 308-314.
Dragović, S.; Onjia, A.; Stanković, S.; Aničin, I.; Bačić, G. *Nucl.Instrum.Meth.* A 2005, 540, 455-463.
Duff, M.C.; Amrhein, C. *Soil Sci. Soc. Am. J.* 1996, 60, 1393-1400.
Eikenberg, J.; Beer, H.; Bajo, S. In Energy, Waste and the Environment: a Geochemical Perspective; Gieré, R.; Stille, P.; Eds.; Geological Society Special Publication 236; The Geological Society: London, 2004, pp. 143-151.
El-Arabi. J. Environ. Radioactiv. 2005, 81, 11-19.
Elejalde, C.; Herranz, M.; Romero, F.; Legards, F. *Water Air Soil Poll.*1996, 89, 23-31.
ESRI, 2009.Environmental Systems Research Institute Inc., Redlands, CA.
Fernandes, H.M.; Lamego Simoes Filho, F.F.; Perez, V.; Franklin, M.R.; Gomiero, L.A. *J. Environ. Radioactiv.* 2006, 88, 140-157.
Fomina, M.; Charnock, J.M.; Hillier, S.; Alvarez, R.; Gadd, G.M. *Environ. Microbiol.* 2007, 9, 1696-1710.
Frissel, M.J.; Noordijk, H.; Van Bergeijk, K.E. In Transfer of radionuclides in natural and semi-natural environment; Desmet, G.; Nassimbeni, P.; Belli, M.; Eds.; Elsevier Applied Science: London, 1990; pp. 40-47.

Gu, B.; Yan, H.; Zhou, P.; Watson, D.B. *Environ. Sci. Technol.* 2005, 39, 5268-5275.

Haas, J.R.; Bailey, E.H.; Purvis, O.W. *Amer. Mineral.* 1998, 83, 1494-1502.

Hancock, G.R.; Grabham, M.K.; Martin, P.; Evans, K.G.; Bollhöfer, A. *Sci. Total Environ.* 2006, 354, 103-119.

Hansen, R.O.; Huntington, G.L. *Soil Sci.* 1969, 108, 257-265.

Harb, S. *Radiat. Prot. Dosim.* 2008, 130, 376-384.

Howard, B.J.; Beresford, N.A.; Hove, K. *Health Phys.* 1991, 61, 715–725.

Howard, B.J.; Strand, P.; Assimakopoulos, P.; Bréchignac, F.; Gascó, C.; Métivier, H.; Moberg, L.; Smith, J.T.; Tamponnet, C.; Trueba, C.; Voigt, G.; Wright, S. In The Radioecology – Ecotoxicology of continental and estuarine environments, ECORAD 2001; Bréchignac, F.; Ed.; Radioprotection – Colloques 37(C1), 2002, pp 1167-1173.

HPS (Health Physics Society).NORM Deposition in the Pulp and Paper Industry; HPS, MacLean, 2000.

Hsi, C.-K.; Langmuir, D. *Geochim.Cosmochim. Ac.* 1985, 49, 1931-1941.

Hu, Q.H.; Weng, J.Q.; Wang, J.S. *J. Environ. Radioactiv.* 2010, 101, 426-437.

IAEA (International Atomic Energy Agency). Applicability of Monitored Natural Attenuation at Radioactively Contaminated Sites; Technical Report Series No. 445; International Atomic Energy Agency: Vienna, 2006.

IAEA (International Atomic Energy Agency). Extent of Environmental Contamination by Naturally Occurring Radioactive Material (NORM) aAnd Technological Options for Mitigation; Technical Report Series No. 419; International Atomic Energy Agency: Vienna, 2003.

IAEA (International Atomic Energy Agency). Radiological Conditions at the Semipalatinsk Test Site, Kazakhstan: Preliminary Assessment and Recommendations for Further Study; Radiological Assessment Reports Series; International Atomic Energy Agency: Vienna, 1998.

IAEA (International Atomic Energy Agency). The Long Term Stabilization of Uranium Mill Tailings; IAEA TECDOC 1403; International Atomic Energy Agency: Vienna, 2004.

Igwe, J.C.; Nnorom, I.C.; Gbaruko, B.C. *Afr. J. Biotechnol.* 2005, 4, 1541–1547.

Ivanovich, M. *Radiochim. Acta* 1994, 64, 81-94.

Jacob, P.; Paretzke, H.G.; Rosenbaum, H.; Zankl, M. *Radiat. Prot. Dosim.* 1986, 14, 299–300.

Janković Mandić, Lj.; Dragović, R.; Dragović, S. J. *Geochem. Explor.*2010, 105, 43-49.

Janković Mandić, Lj.; Dragović, S. *Radiat. Prot. Dosim*. 2010, 140, 369-377.
Kabata-Pendias, A.; Pendias, H. *Soil Sci*. 2005, 170, 743-757.
Karakelle, B.; Öztűrk, N; Köse, A.; Varinlioğlu, A., Erkol, A.Y.; Yilmaz, F. *J. Radioanal. Chem*. 2002, 254, 649-651.
Kemski, J.; Klingel, R.; Schneiders, H.; Siehl, A.; Wiegand J. *Radiat. Prot. Dosim*. 1992, 45, 235-239.
Klement A.W. In Handbook of Environmental Radiation; Ed.; CRC Press: Boca Raton, FL, 1982.
Klimkhanner, G.P.; Palmer, M.R. *Geochim.Cosmochim. Ac*. 1991, 55, 1799-1806.
Kocher, D.C.; Sjoreen, A.L. *Health Phys*. 1985, 48, 193–205.
Koch-Steindl, H.; Pröhl, G. *Radiat. Environ. Bioph*. 2001, 40, 93-104.
Komatina, M.; Komatina, S.M.In Groundwater in the Urban Environment. Selected City Profiles; Chilton, J.; Ed.; Balkema: Rotterdam, 1999; pp 317–322.
Konoplev, A.; Kaminski, S.; Klemt, E.; Konopleva, I.; Miller, R.; Zibold, G. *J. Environ. Radioactiv*. 2002, 58, 1-11.
Korobova, E.; Chizhikova, N.P.; Linnik, V.G. *Euras. Soil Sci*. 2007, 40, 367-379.
Landa, E.R. Isolation of Uranium Mill Tailings and Component Radionuclides from the Biosphere: Some Earth Science Perspectives; U.S. Government Printing Office: Washington, DC, 1980.
Leung, K. C.; Lau, S. Y.; Poon, C. B. *J. Environ. Radioactiv*. 1990, 11, 279–290.
Livens, F.R.; Rimmer, D.L. *Soil Use Manage*. 1988, 4, 63-69.
Livens, F.R.; Horrill, A.D.; Singleton, D.L. *Estuar.Coast. Shelf Sci.* 1994, **38,** 479–489.
Lozano, R.L.; Hernández-Ceballos, M.A.; Adame, J.A.; Casas-Ruíz, M.; Sorribas, M. *Environ. Intern*. 2011, 37, 1259-1264.
Mandarino, F. Fleischer's Glossary of Mineral Species; Mineralogical Record: Tuscon, AZ, 1999.
Manalopoulou, M.; Vagena, E.; Stoulos, S.; Ioannidou, A.; Papastefanou, C. *J. Environ. Radioact*. 2011, 102, 796-797.
Martin, A.; Meade, S.; Wade, B.O. Materials Containing Natural Radionuclides in Enhanced Concentrations, Rep. EUR 17625EN; Commission of the European Communities: Luxembourg, 1997; pp 88.
Mason, B.; Moore, C. B. Principles of Geochemistry; John Wiley & Sons, Inc.: New York, NY, 1982; pp344.

Molinari, J.; Snodgrass, W. J. In The Environmental Behaviour of Radium; IAEA Technical Reports Series No 310; International Atomic Energy Agency: Vienna, 1990; Vol. 1, pp 11-56.

Mortvedt, J.J. *J. Environ. Qual.* 1994, 23, 643-650.

Napier, B.A.; Reed, P.R. Alternative Conceptual Models for Assessing Food Chain Pathways in Biosphere Models, Pacific Northwest National Laboratory; U.S. Nuclear Regulatory Commission Office of Nuclear Regulatory Research: Washington, DC, 2006.

NCRP (National Council on Radiation Protection and Measurements). Exposure of the Population of the United States and Canada from Natural Background Radiation; NCRP Report No. 94; NCRP: Bethesda, MD, 1987.

Nykyri, M.; Nordman, H.; Marcos, N.; Löfman, J.; Poteri, A.; Kautojärvi, A. Radionuclide Release and Transport – RNT-2008; Report 2008-06; Posiva Oy: Olkiluoto, 2008.

Othman, I.; Al-Masri, M.S. *Appl. Radiat. Isotopes* 2007, 65, 131-141.

Oyedele, J.A.; Shimboyo, S.; Sitoka, S.; Gaoseb, F. *Nucl.Instrum.Meth.* A 2010, 619, 467-469.

Papastefanou, C.; Manolopoulou, M.; Charalambous, S. *Radiat. Prot. Dosim.* 1988, 24, 439-443.

Papp, Z.; Dezső, Z.; Daróczy, S. *J. Environ. Radioactiv.* 2002, 59, 191-205.

Payne, T.E.; Davis, J.A.; Lumpkin, G.R.; Chisari, R.; Waite, T.D. *Appl. Clay Sci.* 2004, 26, 151-162.

Pfingsten, W.; Hadermann, J.; Perrochet, P. *J. Contam.Hydrol.* 2001, 47, 349-363.

Pittauerová, D.; Hettwig, B.; Fischer, H.W. *J. Environ. Radioact.* 2011, 102, 877-880.

Prakash, V.; Narayana, Z., Siddappa, K. J. *Radioanal.Nucl. Ch.* 2007, 274, 367-372.

Raabe, O.G. Studies of the solubility of naturally-occurring radionuclides in petroleum scale, NORM/NARM: Regulation and Risk Assessment; Health Physics Society: Scottsdale, AZ, 1996; 121-128.

Ragnarsdottir, K.V.; Charlet, L. In Environmental Mineralogy: Microbial Interactions, Anthropogenic Influences, Contaminated Land and Waste Management; Cotter-Howels, K.V.; Batchelder, J.D.; Campbell, J.; Valsami-Jones, E.; Eds.; Mineralogical Society of Great Britain and Ireland: London, 2000, pp 333-377.

Rajaretnam, G.; Spitz, H.B. *Health Phys.* 2000, 78, 191-198.

Ramola, R.C.; Choubey, V.M.; Prasad, G.; Gusain, G.S.; Tosheva, Z.; Kies, A. *Curr. Sci.* 2011, 100, 906-914.

Ritz, K. In Fungi in Biogeochemical Cycles; Gadd, G.M.; Ed.; Cambridge University Press: Cambridge, 2006, pp 51-73.

Rivas, M. C. Interactions between Soil Uranium Contamination and Fertilization with N, P and S on the Uranium Content and Uptake of Corn, Sunflower and Beans, and Soil Microbiological Parameters (Special Issue); Landbauforschung Völkenrode–FAL Agricultural Research: Braunschweig, 2005.

Rogers, J.J.W.; Adams, J.A.S. In Handbook of Geochemistry; Wedepohl, K. H.; Ed.; Springer-Verlag: Heidelberg, 1969; pp 92.

Rood, A.S.; Voillequé, P.G.; Rope, S.K.; Grogan, H.A.; Till, J.E. *J. Environ. Radioactiv.* 2008, 99, 1258-1278.

Rouni, P.K.; Petropoulos, N.P.; Anagnostakis, M.J.; Hinis, E.P.; Simopoulos, S.E. *Sci. Total Environ.* 2001, 272, 261-272.

Sadusky, M.C.; Sparks, D.L.; Noll, M.R.; Hendricks, G.J. *Soil Sci. Soc. Am. J.* 1987, 51, 1460-1465.

Sanchez, A.L.; Horrill, A.D.; Howard, B.J.; Singleton, D.L.; Mondon, K. *Water Air Soil Poll.* 1998, **106,** 403–424.

Sanchez, A.L.; Wright, S.M.; Smolders, E.; Naylor, C.; Stevens, P.A.; Kennedy, V.H.; Dodd, B.A.; Singleton, D.L.; Barnett, C.L. *Environ. Sci. Technol.* 1999, 33, 2752-2757.

Sawhney, B.L. *Clay. Clay Miner.* 1972, 20, 93-100.

SCK-CEN, Synoptique de la problématique NORM dans l'industrie belge, NRG en AV-Controlatom R-3775; SCK-CEN: Mol, 2003.

Sedlet, J. In Treatise on Analytical Chemistry; Kolthoff, I.M.; Elving, P.J.; Eds.; Wiley: New York, 1966; Vol. 4, pp 219-366.

Selenska-Pobell, S.; Flemming, K.; Tzvetelina, T.; Raff, J.; Schnorpfeil, M.; Geissler, A. In Uranium in the Aquatic Environment; Merkel, B.; Planer-Friedrich, B.; Wolkersdorfer, C.; Eds.; Springer: Heidelberg, 2002; pp 455-465.

Shanbag, P.; Choppin, G. J. Inorg.Nucl. Chem. 1981, 43, 3369-3372.

Sheppard, M.I. The environmental behaviour of uranium and thorium; AECL-6795; Atomic Energy of Canada Ltd.: Pinawa, 1980.

Sheppard, S.C.; Evenden, W.G..*J. Environ. Radioactiv.* 1988, 8, 255-285.

Sheppard, S.C.; Evenden, W.G.; Pollock, R.J. *Can. J. Soil. Sci.* 1989, 69, 751-767.

Shiva Prasad, N.G.; Nagaiah, N.; Ashok, G.V.; Karunakara, N. *Health Phys.* 2008, 94, 264-271.

Smith, I.M. Trace Elements from Coal Combustion: Emissions; Rep. IEA CR/01; IEA Coal Research: London, 1987.

Stamberg, K.; Venkatesan, K.; Vasudeva Rau, P. *Colloid.Surface*. 2003, 221, 149-162.

Tahir, S.N.; Jamil, K.; Zaidi, J.H.; Arif, M. *Radiat. Prot. Dosim*. 2006, 118, 345-351.

Tieh, T. T.; Ledger, E. B.; Rowe, M. W. *Chem. Geol.* 1980, 29, 227-248.

Torsvik, V.; Øvreås, L. In Modern Soil Microbiology; Van Elsas, J.; Jansson, J.K.; Trevores, J.T.; Eds.; CRC Press: Baton Rouge, FL, 2007.

Turner, M.; Rudin, M.; Cizdziel, J.; Hodge, V.*Environ. Pollut*. 2003, 125, 193-203.

UNSCEAR (United Nations Scientific Committee on the Effects of Atomic Radiation). Sources and effects of ionizing radiation, UNSCEAR 2008 Report to the General Assembly with Scientific Annexes; United Nations Scientific Committee on the Effects of Atomic Radiation: New York, NY; 2010; Vol. I.

USDOE/NV (United States Department of Energy, Nevada Operations Office), United States Nuclear Tests, July 1945 through September 1992; USDOE/NV: Las Vegas, 1992.

USEPA (United States Environmental Protection Agency). Understanding Variation on Partition Coefficient, *K*d, Values: II. Review of Geochemistry and Available *K*d Values for Cadmium, Caesium, Chromium, Lead, Plutonium, Radon, Strontium, Thorium, Tritium (^3H) and Uranium; Report 402-R-99-004B; USEPA: Washington, DC, 1999.

USGS (United States Geological Survey), Natural Radioactivity in the Environment; USGS Factsheet; USGS: Denver, CO, 1994.

Van den Bygaart, A.J.; Protz, R.; McCabe, D.C. *Can. J. Soil Sci*. 1999, 79, 161-171.

Van Netten, C.; Morley, D.R. *Arch. Environ. Health* 1983, 38, 172-175.

Van Schmus, W.R. In Global Earth Physics: A Handbook of Physical Constants; Ahrens, T.J.; Ed.; AGU Reference Shelf 1; American Geophysical Union: Washington, DC, 1995; pp 283–291.

Vandenhove, H.; Sweeck, L.; Mallants, D.; Vanmarcke, H.; Aitkulov, A.; Sadyrov, O.; Savosin, M.; Tolongutov, B.; Mirzachev, M.; Clerc, J.J.; Quarch, H.; Aitaliev, A. *J. Environ. Radioactiv*. 2006, 88, 118-139.

Vandenhove, H.; Zeeveaert, A.; Bousher, D.; Jackson, D.; Lambers, B.; Hedemann Jensen, P. Investigation of a Possible Basis for a Common Approach with Regard to the Restoration of Areas Affected by Lasting Radiation Exposure as a Result of Past or Old Practice or Work Activity –

CARE; Final Report to the European Commission, Contract 96-ET-006, Radiation Protection 115; Office for Official Publication by the European Communities: Luxembourg, 2000; pp 242.

Volesky, B. Biosorption of Heavy Metals; CRC Press: Boca Raton, FL, 1990.

Waite, T. D.; Davis, J. A.; Payne, T. E.; Waychunas, G. A.; Xu, N..*Geochim.Cosmochim. Ac.* 1994, 58, 5465-5478.

WISMUT, Sanierungskonzept Standort Dresden-Gittersee; Wismut GmbH: Chemnitz; 1992.

Wright, S.; Smith, J.T.; Beresford, N.A.; Scott, W.A. *Radiat. Environ. Bioph.* 2003, 42, 41-47.

Wysocka, M.; Skowronek, J. In Nuclear Techniques in the Exploration and Exploitation of Energy and Mineral Resources; International Atomic Energy Agency: Vienna, 1991; pp 457-462.

Zhang, P.-C.; Krumhansl, J.L.; Brady, P.V. In Geochemistry of Soil Radionuclides; Zhang, P.-C.; Brady, P.; Eds.; Soil Science Society of America: Madison, WI, 2002.

Zhiyanski, M.; Bech, J.; Sokolovska, M.; Lucot, E.; Bech, J.; Badot, P-M.*J. Geochem. Explor*. 2008, 96, 256-266.

Zhu, B.; Alva, A.K. *Soil Sci. Soc. Am. J.* 1993, 57, 350-355.

In: Radiation Exposure
Editors: D. Balenovic and E. Stimac

ISBN 978-1-62100-869-9

Chapter 2

RADIATION EXPOSURE AND PROTECTION IN CONVENTIONAL PEDIATRIC RADIOLOGY OF THORAX

Vera Garaj-Vrhovac[1*], Goran Gajski[1], Đurđica Milković[2], Mária Ranogajec-Komor[3], Saveta Miljanić[3], Natko Beck[4], Željka Knežević[3], and Marko Gerić[1]

Institute for Medical Research and Occupational Health, Mutagenesis Unit, 10000 Zagreb, Croatia[1]

Children Hospital Srebrnjak, Radiology and Ultrasound Unit, 10000 Zagreb, Croatia[2]

Institute Ruđer Bošković, *Laboratory for Radiation Chemistry and Dosimetry*, 10000 Zagreb, Croatia[3]

University Hospital Sveti Duh, Radiology Unit, 10000 Zagreb, Croatia[4]

ABSTRACT

Radiation protection depends on many factors that can be divided on the ground of technical and human factors that are both equally relevant for the radiation protection. To provide the best care for the young patients the effects of radiation should be minimized in accordance with ALARA (As Low As Reasonably Achievable) principle. Chest

[*]Correspondence: V. Garaj-Vrhovac, Institute for Medical Research and Occupational Health, Mutagenesis Unit, Ksaverska cesta 2, 10000 Zagreb, E-mail: vgaraj@imi.hr.

radiograms represent the basic radiological examination of thorax as they provide diagnostic information on the lung and heart status, and the skeletal system. To provide the best care for the young patients the effects of ionizing radiation should be minimized. Dosimetry is an indispensable part of every investigation related to the optimization and improvement of radiation protection. This study is dealing with the significance of different dosimetry systems that are applied in X-rays diagnostics, determination of doses received by patients and with the role of human factor during the radiological examination. Also the impact of ionizing radiation on cytogenetic status of children exposed to X-rays, since understanding the risks of low doses of radiation is an important aspect in the risk benefit analysis in pediatric populations. The study comprised child population that had a clinical indication for chest X-ray examination. Absorbed dose measurements were conducted on the forehead, thyroid gland, gonads, chest and back of the patients as well as on the phantom for radiological examinations of thorax. In this work doses were measured using various thermoluminescence (TL) and recently radiophotoluminescence (RPL) dosimetry systems. RPL and TL dosimeters were placed on the patients at the entrance of the beam in the centre of the X-ray field to measure the entrance dose. For comparison, during chest radiography, the dose in anthropomorphic child phantom was measured with TL dosimeters. The results on patients with both dosimetry systems showed a good correlation of measured dose. Also dose values on phantom were comparable with the results obtained on patients. If all technical malfunctions are excluded equal responsibility is on radiology technicians that are responsible for the patient's dose. Depending on their education and experience, technicians perform X-ray examination with various end results. In this study doses measured by two technicians were in a good agreement, for the third one doses were significantly higher. Following the analysis of the dosimetry results and the education of the personal the doses measured on children for all of them were in very good agreement and lower than before education. For the cytogenetic monitoring comet assay and micronucleus test were applied as a sensitive biomarkers for the detection of DNA damage and chromosomal instability after irradiation. Results demonstrated that there is significant potential of inducing genetic instability with even the low level diagnostic X-ray procedures. Radiological examination of children is extremely significant since it stands for the sole and most accurate investigation method. Radiation protection depends on many factors and a suitable cooperation of all medical doctors especially radiologists would be recommended. In conclusion, this work shows the importance of continues education and good teamwork for dose reduction to minimize the potential damage and maintain the optimal benefit for the young patients.

Keywords: radiation exposure, radiation protection, pediatric radiology, dosimetry systems, TLD, RPL, genomic instability, comet assay, micronucleus assay, ALARA

1. Introduction

Ever since the great invention of Wilhelm Conrad Röntgen in the late 19 century medicine uses radiation for diagnostic purposes and later also for therapeutical applications (Röntgen 1898). The most common use of ionizing radiation is in the field of diagnostic radiology, where X-rays are used to give a doctor a view into the patient's condition. Radionuclids are also used in diagnostic procedures in nuclear medicine. Applied in a different way radiation is used in medicine for therapeutical means, where high energy X-rays and radionuclides are used to suppress the growth of tumors (Brugmans et al. 2002; Mettler et al. 2008). Interventional radiology uses radiation based procedures for image acquisition so that a therapeutical, often life saving, intervention inside the human body is made possible (Sacks et al. 2003).

1.1. X-Ray in Medicine

Diagnostic radiology uses external radiation to create images of the body, its organs and gives doctors an insight into a patient's condition to help in the diagnostic pathway and treatment. When indicating a diagnostic procedure in which radiation is involved a clinician always has to have in mind the benefits and the disadvantages of procedure. As radiation is potentially harmful for cells and the genetic material, one must always secure that procedures which involve radiation are performed at the highest level possible to have the least harmful effect on the patient (Slovis 2002).

Diagnostic radiology that is radiation dependent is usually divided into conventional or plain radiography, fluoroscopy and angiography, and computed tomography (Daffner 2007). Conventional radiography uses a beam of X-rays that are generated in the X-ray tube. The beams are aimed at the patient, better said at the zone of interest, for example thoracic region, bones etc. The rays then hit the undeveloped film or today also sensor plate from which an image is obtained. Today digital radiography is slowly replacing the film by converting signals from the sensor plate into digital signals then viewable on a computer screen. Plain radiography is the most often diagnostic radiological procedure thanks to its speed, low dose, low cost and availability (Als-Nielsen & McMorrow 2001).

Fluoroscopy and angiography are based on the same principles as plain radiography yet the picture is obtained by a fluorescent screen and image intensifier tube and then viewed in real time on a TV screen. This allows doctors to view the motility of certain organs in the body as well as their function. Contrast agents that can be applied oral, intravenous/arterial/lymphatic, per rectum, per urethral, per vaginal, laryngo-tracheal, intraarticular and intraspinal help to delineate the aimed system's anatomy (Thomsen & Webb 2009). These procedures have a higher dose per examination and in interventional radiology depending on the length of the procedure ergo the radiation exposure the doses are the highest in diagnostic radiology, not only for the patient but also for the team performing the examination (Williams 1997).

Computed tomography (CT) uses X-rays and detectors to obtain images that are produced by computer algorithms and then viewed on a computer screen. The images are obtained in only one plane, the so called axial plane. Two other planes are then computed, the coronal and the sagittal plane. The progress in the CT industry led to better and faster machines, the Spiral Multidetector CT using 8, 16, 64 detectors that give a better image in a shorter time, with more details and anatomical delineation, also three dimensional images of different systems or organs and have so become one of the best diagnostic procedures especially in urgent medicine as they provide accurate information very fast. CT exams can also be combined with different contrast agents to provide more information. Depending on the size of the field that is viewed, the distance between the each tomogram the dose for the patient varies from medium to sometimes higher dose (Taguchi & Anno 2000; Brix et al. 2003; Honnef et al. 2004; Coles et al. 2006; Einstein et al. 2007).

1.2. Dosimetry in Medicine

Doses from diagnostic radiology procedures are small and usually do not approach thresholds for deterministic effects. Exceptions are found for interventional procedures that may involve high doses to the patient's skin. But even ignoring the high doses found for interventional procedures, it needs to be realized that the greatest source of exposure of the population to artificial ionizing radiation is from diagnostic radiology. These doses should therefore be accurately determined (especially in pediatric radiology) in order to maintain a reasonable balance between image quality and patient exposure. Dosimetric methods should be used that ensure appropriate levels of accuracy and long term stability (IAEA 2007).

The ionization chamber is the main device used for diagnostic radiology measurements, but for the dose measurements on patients the passive solid state detectors are very important, among them the most popular are thermoluminescence (TL) detectors and recently radiophotoluminescent (RPL) detectors. In TL and RPL dosimetry, the relationship between the relevant signal and the dose to be measured must be determined by calibration. In diagnostic radiology, the specification of radiation qualities is important as the response of all dosimeters depends at least to a certain extent on spectral distribution of the X-rays employed. Radiation qualities are usually specified in terms of the X-ray tube voltage first and possibly second half value layer (HVL). The dose on patients depends mainly on the tube voltage and quantity of charge (IAEA 2007).

1.2.1. Thermoluminescence (TL) Dosimetry

Thermoluminescence is the emission of light when a material that has been exposed to ionizing radiation is heated. This light result from the release of electrons that were excited and trapped when the material was irradiated, and the amount of light released is directly related to the radiation dose received by the material. In using this phenomenon for dosimetry, the TL material is observed by a photomultiplier or other light sensitive device during the heating process. A plot of the luminescent light output against temperature is called the "glow curve". The area under the glow curve is used as a measure of dose. The TL material is discharged by the reading process and some materials are ready to register a new exposure and some must be annealed before reissue.

The mechanism of thermoluminescence is complex, and although general theoretical models have been postulated, each TL phosphor is unique, and models which correspond to particular materials display very different characteristics.

TL dosimeters most commonly used in medical applications are based on lithium fluoride doped with magnesium and titanium (LiF:Mg,Ti) and with magnesium, copper and phosphor (LiF:Mg,Cu,P) but other materials such as LiF:Mg,Cu,Si, $Li_2B_4O_7$:Mn, $CaSO_4$:Dy and CaF_2:Mn have also been used. Some basic characteristics of these materials are given in many publications (IAEA 1999; Lee et al. 2006; IAEA 2007).

There are many factors that influence the final result of a thermoluminescence measurement. These include factors related to the performance of the instrument and those related to procedures of dosimeter preparation and handling. Extensive descriptions of various TL materials and procedures for the preparation, handling and evaluation can be found in literature (Horowitz 1984; McKeever et al. 1994). A review of the application of a thermoluminescent method for dosimetry in medicine was recently prepared by Ranogajec-Komor (2004) as well as in diagnostic radiology was done by Zoetelief et al. (2000). Advantages of TLD include the existence of nearly tissue equivalent TL materials, sufficiently high sensitivity and accuracy, commercial availability of small sized detectors, availability of materials with excellent long term stability, ease of processing, reusability and linearity of response with dose and dose rate over a large range.

1.2.2. Radiophotoluminescence (RPL) Dosimetry

Radiophotoluminescence is defined as the luminescence observed when some minerals or glasses, having been exposed to ionizing radiation, are subsequently exposed to ultraviolet light. The one selected as the most economical and practical for large-scale manufacture was a silver-activated phosphate glass introduced by Schulman & Etzel (1953). Photoluminescence is based on the formation of stable induced luminescent centers in silver doped phosphate glasses when they are exposed to ionizing radiation. When the glasses are subsequently exposed to ultraviolet radiation, orange fluorescent light is emitted with an intensity that is linearly related to the absorbed dose from the ionizing radiation. Unlike thermoluminescence, the effect of the ionizing radiation – the centers – do not disappear by the normal reading process, and are extremely stable, so that fading at room temperature is negligible over a period of several years and the dose information can be obtained at any time during long term dose accumulation (IAEA 1999).

Phosphate glasses can be produced on a large scale with good reproducibility and constant sensitivity. Thus, calibration of individual detectors is not needed. The application of commercially available pulsed ultraviolet laser readers reduces the "pre-dose" – the apparent reading from unirradiated glasses - to a value of about 10 μSv; that giving capability of measuring doses as low as several μSv (Piesch et al. 1990). Because of the high atomic number of glass materials, energy compensation filters have to be used, giving an energy dependence within ± 15% for photon energies above 15 keV (Burgkhardt et al. 1983).

The advantages of photoluminescent dosimeters include permanent and long term integration of dose information, easy of the reading and processing, good accuracy, negligible fading, high uniformity and possibility of repeating a dosimeter reading if necessary.

Properties of different types of RPL glasses prepared by then were compared in the review paper by Piesch (1972). Among others it is important to point out the work by Yokota & Muto (1971) who prepared several types with improved characteristics, among them type FD-7 is still a basis for different types of RPL glass dosimeters.

In the early eighties new flat glass dosimeters developed at Kernforschungszentrum Karlsruhe extended the low photon energy range from 40 keV to about 10 keV (Burgkhardt et al. 1983). The flat phosphate glass dosimeter SC-1 (with glass element FD-7) was optimized with respect to the design of the capsule and energy compensation filter.

The new generation of automatic readers, FGD-200 series, enables continuously reading of 20 dosimeters; reading time is 10 s or less per dosimeter and dose range is 1 μGy (Sv) – 10 Gy (Sv) (TGC 2003). Characteristics of the SC-1/FGD-200 system were tested by different authors (Hsu et al. 2006; Miljanić et al. 2008; Ranogajec-Komor et al. 2008; Lee et al. 2009).

SC-1 is used for different applications in personal and area monitoring (Piesch et al. 1994; Piesch & Burghardt 1994; Burghardt et al. 1996a; Burghardt et al. 1996b). It is also used for patient dosimetry (Milković et al. 2009) but these applications are limited due to large dimensions of SC-1 dosimeters.

For the applications in radiobiology and patient dosimetry in medicine the small dimensions of dosimeters are of great importance. Recently Asahi Techno Glass Corporation (TGC 2004) developed a new system "Small Element System Dose Ace" using very small glass elements GD-300 series for various applications primary in medicine. There are two types of glass rods, both have diameter of 1.5 mm, but the length is different, 8.5 and 12 mm. Every type of rod is placed in a plastic holder which can have an energy compensation Sn filter or not. Types GD-351 (8.5 mm rod length) and GD-352 (12 mm rod length) contain Sn filter and are suitable for medium and low X-rays as used in diagnostic radiology. Details about measurement technique and characteristics of the system without Sn filter were described by Tsuda (2000), Araki et al. (2003), Araki et al. (2004), Nose et al. (2005) and Perks et al. (2005). The experimental data using Dose Ace system (with Sn filter) for patient doses in diagnostic radiology are still scarce. Akahane et al. (2003)

measured neonates patients doses in CT examinations, Ono et al. (2005) measured digital and conventional chest radiographic doses using a thorax phantom, Ohuchi et al. (2005) measured skin doses of patients in interventional radiology procedures. Two papers compare dosimetric characteristics of Dose Ace with filters and TL dosimeters, Zhuo et al. (2007) (GD-351 and LiF: Cu, Mg,P) and Knežević et al. (2011) (GD-352 and LiF:Cu,Mg,P). Authors in both papers found that the examined properties for RPL and TL dosimeters are very similar but uniformity (Zhuo et al. 2007; Knežević et al. 2011) and fading effect (Zhuo et al. 2007) are considerable better for RPL.

1.3. Biodosimetry

In addition to physical dosimetry, it also became useful to conduct biological dosimetry or biodosimetry, with the aim of individual dose assessment. In that manner, comet assay and micronucleus assay have both become useful biomarkers for radiation exposure. Many studies have shown that the number of radiation-induced DNA strand breaks and micronuclei is strongly correlated with dose and quality of radiation. Both the comet assay and the micronucleus assay have become, in the last years, thoroughly validated and standardized techniques to evaluate *in vivo* radiation exposure of occupational, medical and accidentally exposed individuals (Tice & Strauss 1995; Garaj-Vrhovac et al. 2002; Maluf 2004; Milković et al. 2009; Olive 2009; Fenech 2010; Sirota & Kuznetsova 2010; Vral et al. 2011).

1.3.1. Comet Assay

The comet assay is a rapid, sensitive and relatively simple method for detecting DNA damage at the level of individual cells (Tice & Strauss 1995; Collins et al. 1997; Collins et al. 2008; Dusinska & Collins 2008; McArt et al. 2009). It combines the simplicity of biochemical techniques for detecting DNA single strand breaks, alkali labile sites, and cross-linking, with the single cell approach typical of cytogenetic assays, and has become one of the standard methods for assessing genome damage with a variety of applications in genotoxicity testing, fundamental research in DNA damage, and repair as well as in population biomonitoring (Collins 2004; Møller 2006; Dhawan et al. 2009; Olive 2009; Valverde & Rojas 2009).

1.3.2. Micronucleus Assay

Micronucleus assay has become one of the best validated methods for measuring chromosome damage in human lymphocytes (Fenech et al. 2003; Fenech 2006; Fenech 2007). Apart from the measurement of micronuclei (MNi) frequency, a biomarker of chromosome breakage and/or whole chromosome loss, it is now also possible to measure other important biomarkers within the micronucleus assay that include nucleoplasmic bridges (NPBs), a biomarker of DNA mis-repair and/or telomere end-fusion representing dicentric and ring chromosomes as well as nuclear buds (NBUDs), a biomarker of elimination of amplified DNA and/or DNA repair complexes (Umegaki & Fenech 2000; Fenech et al. 2003; Thomas et al. 2003; Fenech 2010; Fenech et al. 2011).

Both assays have become the standard methods for assessing genome damage with a variety of applications in genotoxicity testing, fundamental research in DNA damage, and repair as well as in population biomonitoring (Fenech et al. 1997; Gajski et al. 2007; Gajski et al. 2008; Garaj-Vrhovac et al. 2008a; Garaj-Vrhovac et al. 2008b; Fenech 2010; Kopjar et al. 2010; Fenech 2011; Garaj-Vrhovac et al. 2011; Vral et al. 2011).

Present chapter describes investigations of the exposure of pediatric patients in X-ray diagnostic of thorax with different dosimetry systems and the assessment of cytogenetic damage induced in this population. For that purpose the genomic instability was evaluated with alkaline version of the comet assay and cytokinesis-block micronucleus (CBMN) assay on human peripheral blood lymphocytes in function of doses measured using TL and RPL dosimetry systems. Such biomonitoring will provide a useful contribution to the characterization of diagnostic X-ray exposure in the youngest population that was assessed since children with pulmonary diseases are at a relatively greater cancer risk as the radiological exposure can be intensive in these patients. The evaluation of early biological effects such as genome damage could contribute to clarify the effect that low dose diagnostic radiation might have and to identify biomarkers of early cell disturbances and DNA damage useful in terms of health risk assessment.

2. Experimental Procedures

Concerning the review character and diversity of results experimental procedures are described partly with results.

2.1. Patients and Blood Sampling

Twenty children (14 female and 6 male), between the ages of 5 and 14 years, were selected randomly, all with pulmonary diseases, such as pneumonia, bronchitis, asthma and suspected tuberculosis to follow the changes of biodosimetry parameters after X-ray examinations. None of the assessed children was exposed to ionizing radiation or steroid therapy for at least the previous 6 months, or antibiotics for the last 3 months. Diagnostic chest X-rays and clinical pathology exams of blood samples were warranted because of the children's health conditions. The present study was performed in accordance with high ethical standards. Blood donors were informed about the aim and the experimental details of the study. All of the parents or legal guardians gave their informed consent and volunteered to donate blood for sampling. Every volunteer completed a standardized questionnaire, designed to obtain relevant details about the current health status, history and lifestyle. No private details on subjects involved in study are or will be highlighted in public. Blood samples were taken from veins in the antecubital fossae prior to, and exactly 30 minutes after X-ray examination. Samples were taken simultaneously during the winter period to minimize different factors during the summer (e.g. solar radiation, air temperature, air ozone level etc.). Peripheral blood samples were collected in heparinized vacutainer tubes (Becton Dickinson, Franklin Lakes, NJ) under aseptic conditions. All blood samples were handled in the same manner. After venipuncture they were randomly coded, cooled to +4°C in the dark, and transferred to the laboratory on ice to minimize DNA repair after irradiation. They were analyzed immediately after the arrival (one hour after collection at the latest) using the alkaline version of comet assay and CBMN assay.

2.2. X-Ray Exposure and Dosimetry

The X-ray examinations of the thorax were carried out in postero-anterial (PA) and lateral X-ray projection on 150 kV Shimadzu CH-200M unit (Tokyo, Japan; Figure 1). The conditions of exposure were chosen the same as were used for the routine diagnostic procedures. The parameters of the irradiation were the following: the voltage of the X-ray tube varied in the range 95 kV - 105 kV, the quantity of charge 2 mAs - 3.6 mAs, the time of irradiation 5 ms - 11 ms. For the organ dose measurements and effective dose estimation child phantom was used. The phantom was an anthropomorphic heterogeneous tissue-equivalent phantom which corresponds to the child age of 1-3 years; the height was 78 cm and the weight 9 kg (Varčenja et al. 1988).

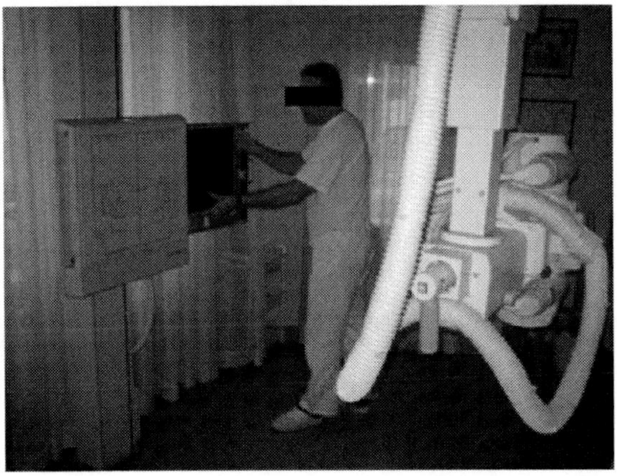

Figure 1. The X-ray examinations of the thorax were carried out in posterior-anterior (PA) projection on 150 kV Shimadzu CH-200M X-ray unit.

The phantom consists of 28 segments with 95 holes for placing the TL detectors (Figure 2). For the phantom measurements the voltage of the X-ray tube was 55 kV - 60 kV, the X-ray tube current, the exposure time and numbers of exposures were 3.2-5mA, 5-12 mS and 20 respectively.

Figure 2. Anthropomorphic child phantom.

Dosimetry systems used in this research were two types of RPL glass dosimetry system (SC-1 and GD-352), two types of TL dosimetry system based on LiF:Mg,Cu,P and TL system based on LiF:Mg,Ti (TLD-100, TLD-700 and MTS-N; Figure 3).

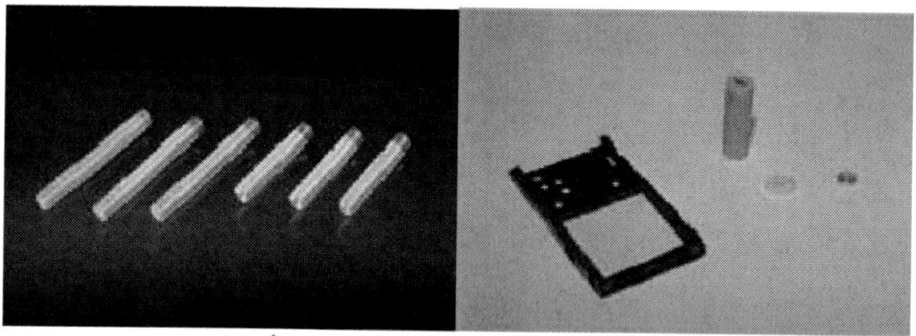

A B

Figure 3. Different types of radiophotoluminescent (A) and thermoluminescence dosimeters (B).

The RPL dosimeters type SC-1 consist of flat silver activated phosphate glass in a dosimeter capsule containing an energy compensation filter. Inside the dosimeter capsule ($30 \times 40 \times 9$ mm^3, ABS plastic, 3 mm thickness), the flat glass ($16 \times 16 \times 1.5$ mm^3) is fixed in stainless steel holder glass card bearing the number (hole code) and is partly shielded by tin filters of 0.75 mm thickness. The RPL dosimeter type GD-352M (dimensions 12 mm length) is a rod type dosimeter placed in a plastic holder which has an energy compensation Sn filter.

In our experiments the full automatic FGD-200 Series dosimeter system with SC-1 glass dosimeters and FGD-1000 semi-automatic system with GD-352M glass rods dosimeters were used. Both systems are based on pulsed nitrogen gas UV laser excitation. Because fluorescence is emitted whenever the laser pulse excites the glass, one glass element can be readout repeatedly many times.

The TL dosimetry systems consisted of two types of high sensitivity LiF:Mg.Cu,P dosimeters and LiF:Mg,Ti dosimeters (TLD-100) that are usually taken as a reference dosimeter. Because in the X-ray diagnostic of thorax in children patients very low doses are expected LiF:Mg,Cu,P dosimeters were chosen due to its high sensitivity, about 100 times more than TLD-100. For the measurements in tissue equivalent anthropomorphic child phantom LiF:Mg,Ti dosimeters were used. The reading of TL dosimeters were carried out using a modified TOLEDO 654 (Vinten Instruments Ltd, Beaconsfield, UK) reader. The reader is connected with a PC containing software which enables the integration of the glow curves with variable integration limits and a detailed analysis of glow curves.

One of the advantages of both RPL and TL dosimetry systems is the possibility of their reuse. Before every irradiation both RPL and TL dosimeters were annealed in special furnaces. Some properties of the used dosimeters as well as some parameters used for the pre-irradiation annealing, preheat and readout are presented in Table 1.

For calibration purposes ^{137}Cs gamma ray source was used. The individual sensitivity of each TL detector was previously determined by irradiation with a dose of on ^{137}Cs while the RPL detectors were calibrated individually by the manufacturer.

In all the experiments 5-10 TL and RPL dosimeters were chosen in order to estimate the calibration factors. TL detectors and RPL detectors placed "free in air" were irradiated with doses of 2 mGy. For background correction 3-5 control dosimeters of each kind were used. During X-ray diagnostics of chest in children surface doses were measured on the forehead, back, sternum, thyroid gland, eye and gonads. In some cases to ensure radiogram quality RPL dosimeters type SC-1 were not placed on chest and on the back.

Table 1. The evaluation parameters of the investigated detectors

Dosimeter	RPL		TLD		
Type	SC-1	GD-352M			
Material	silver activated phosphate glass		LiF:Mg,Cu,P		LiF:Mg,Ti
Origin	Japan		China	USA	USA
Size (mm)	15x15	12 mm rod length	ϕ 4.5x0.8	3x3x0.5	3x3x0.9
Preirradiation annealing Temperature (°C) Time (min)	400 60	400 20	240 10		400 +100 60 + 120
Preheat Temperature (°C) Time (min)	70 60	70 30	100 20		100 20
Readout Temperature (°C) Time (s)	UV excitation		220 30		270 35

2.3. Biodosimetry

2.3.1 Alkaline Comet (SCGE) Assay

To estimate the level of primary DNA damage after exposure to diagnostic X-rays the comet assay was carried out under alkaline conditions, as previously described by Singh et al. (1988). Fully frosted slides were covered with 1% normal melting point (NMP) agarose (Sigma, St Louis, MO). After solidification, the gel was scraped off the slide. The slides were then coated with 0.6% NMP agarose. When this layer had solidified, a second layer, containing the whole blood sample mixed with 0.5% low melting point (LMP) agarose (Sigma) was placed on the slides. After 10 min of solidification on ice, slides were covered with 0.5% LMP agarose. Slides were then immersed for 1 h in ice-cold freshly prepared lysis solution [2.5 M NaCl, 100 mM disodium EDTA, 10 mM Tris–HCl, 1% sodium sarcosinate (Sigma), pH 10] with 1% Triton X-100 (Sigma) and 10% dimethyl sulfoxide (Kemika, Zagreb, Croatia) to lyse the cells and allow DNA unfolding. The slides were then placed on a horizontal gel electrophoresis tank, facing the anode. The unit was filled with fresh electrophoresis buffer (300 mM NaOH, 1 mM disodium EDTA, pH 13.0) and the slides were placed in this alkaline buffer for 20 min. to allow DNA unwinding and expression of alkali-labile sites. Denaturation and electrophoresis were performed at 4°C under dim light. Electrophoresis was carried out for 20 min at 25 V (300 mA) (20 min at 1 V/cm). After electrophoresis, the slides were rinsed gently three times with neutralization buffer (0.4 M Tris–HCl, pH 7.5) to remove excess alkali and detergents. Each slide was stained with ethidium bromide and covered with a coverslip. Slides were stored at 4°C in sealed boxes until analyzed. To prevent additional DNA damage, handling blood samples and all the steps included in the preparation of the slides for the comet assay were conducted under yellow light or in the dark. One hundred randomly-captured comets from each slide were examined at 250× magnification using an epifluorescence microscope (Zeiss, Oberkochen, Germany) connected through a camera to an image analysis system (Comet Assay II; Perceptive Instruments Ltd, Haverhill, Suffolk, UK). An automated image analysis system was used to acquire images, compute the integrated intensity profile for each cell, estimate the comet cell components and evaluate the range of derived parameters. Comets were randomly captured at a constant depth of the gel, avoiding the edges of the gel, occasional dead cells, and superimposed comets or comets without distinct head ("clouds", "hedgehogs", or "ghost cells"). To avoid potential variability, one well-trained scorer performed all scorings of comets. To quantify DNA damage, the following comet parameters were evaluated: tail length, tail intensity (% of DNA), and tail moment. Tail length (i.e. the length of DNA migration) is presented in micrometers (μm). It was calculated from the center of the cell. Tail intensity is defined as the percentage of fluorescence migrated in the comet tail. Tail moment was calculated as (tail length x % of DNA in tail)/100.

2.3.2. Cytokinesis-Block Micronucleus (CBMN) Assay

Lymphocyte cultures for the CBMN assay after exposure to diagnostic X-rays were incubated according to standard protocol as described by Fenech & Morley (1985). Cells were incubated in a Euroclone medium (Chromosome kit P, Euroclone, Milano, Italy) at 37 °C in an atmosphere of 5% CO_2. Cytochalasin-B (Cyt-B; Sigma, St. Louis, MO) at the final concentration of 6 μg/ml was added to each sample on the 44th hour of incubation, to accumulate cells that had divided once, and the cells were harvested after another 28 hours. After adding saline, cells were fixed with a cold fixative methanol:acetic acid (3:1, v/v). The fixation step was repeated twice; the cells were resuspended in a small volume of fixative solution and dropped onto clean slides. Finally, they were stained for 10 min in a 5% aqueous solution of Giemsa dye (Merck, Darmstadt, Germany). Every subject was analyzed for the total number of MNi, NPBs and NBUDs per 1000 binucleated (BN) cells according to criteria previously published by Fenech et al. (2003). Only BN cells with well-preserved cytoplasm were scored for analysis. Microscopic analysis was done under optical microscope with a final magnification 400× (Olympus CX41, Tokyo, Japan). Scoring of all the samples was performed blind to the exposure status of the subject.

2.3.3. Statistics

Measured parameters were evaluated using Statistica 5.0 package (StaSoft, Tulsa, OK). Multiple comparisons between groups were done by means of ANOVA. Analysis of differences was done by non-parametric Mann-Whitney U test. Statistical significance was defined as $P<0.05$ in all analyses.

3. Results and Discussion

3.1. Determination of Pediatric Patient's Exposure in X- Ray Diagnostic of Thorax with Various Solid State Dosimeters

Exposure of children patients at thorax X-ray examinations were studied during longer period. In our research work, full attention was given to the improvement of TL dosimetric systems, their characterization and calibration in Secondary Standard Dosimetric Laboratories (SSDL) as well as the enhancing of their accuracy (Ranogajec-Komor & Osvay 1986; Ranogajec-Komor et al. 1993; Miljanić et al. 2002; Ranogajec-Komor 2004; Knežević et al. 2010). On the basis of their characteristics such as linearity, sensitivity, reproducibility, fading, energy dependence LiF:Mg,Ti and LiF:Mg,Cu,P systems were chosen for dosimetric measurements on patients. Various RPL systems were also characterized and recently successfully applied for diagnostic measurements (Ranogajec-Komor et al. 2008; Knežević et al. 2011).

The dose distribution on the body surface of children was determined during chest radiography as function of the age of patients which means also change of irradiation conditions (Milković et al. 1991; Ranogajec-Komor et al. 1993). Even the X-ray facility differs depending on the age. The results obtained on 50 patients divided in 5 age groups at postero-anterial and lateral X-ray examination of thorax on various body parts showed large standard deviation (SD; Table 2).

The high SD (up to 100% even within the same age group) indicates the real situation during routine examinations. The distance between patient and X-ray tube, the inclination of the tube, the tube voltage and current as well as the body absorption and radiation scattering varies according to patient's weight and height. For comparison the SD of the same detectors were ±4% after irradiation with 80-83 keV X-ray energies in various SSDL, ±21% in the routine irradiation filed without patients. The large variation of the dose indicates the need of highly statistical dose measurements directly on the patient even in case where precise phantom measurements were made in advance. For comparison the doses in anthropomorphic child phantom were measured with TL dosimeters during chest radiography (Ranogajec-Komor et al. 1997; Milković 1998). The results obtained on the surface of the phantom at different body parts are shown in Table 3.

Table 2. Mean value of the surface dose measured on children during X-ray diagnostics of chest.

Age (years)	Dose (mSv)					
	Back	Sternum	Thyroid	Right armpit	Right eye	Right gonad
0-3	0.12	0.07	0.08	0.05	0.08	0.07
3-6	0.17	0.15	0.12	0.10	0.05	0.04
6-9	0.31	0.29	0.27	0.13	0.04	0.02
9-12	0.36	0.26	016	0.17	0.03	0.02
>12	0.33	0.25	0.20	0.18	0.07	0.03
Mean	0.26	0.20	0.17	0.13	0.05	0.04
SD	0.11	0.09	0.05	0.05	0.02	0.02

Table 3. Dose on the surface of children phantom during postero-anterial (PA) examination of thorax.

Cycle	Dose (mSv)				
	Back	Sternum	Thyroid	Right eye	Right gonad
1.	0.91	0.16	0.11	0.02	0.00
2.	0.90	0.04	0.14	0.01	0.00
Mean	0.91	0.10	0.13	0.02	0.00

Table 4. The surface doses measured on 5 patients during postero-anterial (PA) X-ray diagnostics of chest.

Dose (mSv)	Back	Sternum	Thyroid	Right eye	Right gonad
Mean value	0.08	0.04	0.02	0	0.01
Standard deviation	0.04	0.04	0.03	0	0.01

*- The average age, height and weight of patient was 3 years, 85 cm and 18 kg, respectively.

**- The potential, current of the X-ray tube, the exposure time and the focus to film distance were 115 kV, 1.8-2.2 mAs, 0.01-0.15 s, 150 cm, respectively.

***- Gonad shields were systematically used.

Dose values on phantom measured in two measurement cycles are comparable to the results measured on patients (Table 4) taking into consideration that on the phantom number of exposition was 20 and the irradiation facilities were in these two cases different (phantom: Specialised Children Hospital, Bratislava, Philips Medio 50 CP-H; patient: Children Hospital Srebrnjak, Zagreb Multix facility with rotary tube). The patient's doses were measured on 10 children with similar weight and height as the phantom.

In the tissue-equivalent anthropomorphic child phantom the organ doses were measured during chest radiography by thermoluminescence dosimetry. The number and position of dosimeters were determined by the construction of the phantom. From the organ doses the effective dose (E) was calculated according:

$$E=\sum w_T D_T$$

where w_T is the tissue weighting factor (ICRP 1991), and D_T is the measured absorbed dose averaged over the organ or tissue T. The effective dose obtained in the phantom was 0.019 mSv.

The effective dose *E* defined by equation was used for radiation risk assessment. The risks of leukemia, lung cancer and genetic damages have been estimated using the risk factors from ICRP 60 and the effective dose values calculated from the measured values. Though the evaluated risks were not alarming (from 3-10 x 10^6 examined children one can get carcinogenic damage as a consequence of X-ray irradiation; Milković 1998) all patient protection measures should be carried out especially in pediatric radiology. The dosimetric measurements have proved also the efficiency of various radiation protection methods (collars, blue and green film programs and screens).

Recently the biological effect of low doses in X-ray diagnostic was studied i.e. the assessment of the level of DNA damage in peripheral blood lymphocytes of children before and after their exposure to the X-rays for chest examination. These data were compared to the measured absorbed dose. Radiation exposure estimates at the different positions on the body measured by TLD are presented in Table 5. The forehead and gonads doses on all patients were practically not traceable (0.00-0.01 mSv). RPL were applied on the thyroid, forehead and gonads. The results of both dosimetry systems were in good agreement.

Table 5. Doses (expressed as $H_p(10)$) measured at different positions on the body with thermoluminescence (TL) dosimeters.

Subject No.	Back (mSv)	Sternum (mSv)	Thyroid (mSv)
1*	-	-	0.04
2	0.11	0.01	0.02
3	0.13	0.02	0.03
4	0.15	0.03	0.03
5	0.28	0.02	0.04
6	0.29	0.03	0.06
7	0.24	0.02	0.03
8	0.24	0.04	0.05
9	0.22	0.02	0.04
10	0.29	0.02	0.04
11	0.20	0.02	0.03
12	0.17	0.02	0.02
13	0.27	0.03	0.04
14	0.19	0.02	0.02
15	0.17	0.02	0.03
16	0.17	0.01	0.02
17	0.22	0.03	0.03
18	0.23	0.02	0.03
19	0.18	0.02	0.03
20	0.20	0.02	0.03

*Dosimeters in case of the first patient were not used in primary beam.

The X-radiation exposure measured on the back of 20 patients varied from 0.11 mSv to 0.29 mSv which is in good agreement with the literature data (Simpson et al. 2008) as with the earlier results shown in Table 1 and 3. X-radiation exposures measured on the sternum were from 0.01 mSv to 0.04 mSv. The dose values are expressed as $H_p(10)$. Thyroid exposures ranging from 0.02 mSv to 0.06 mSv were in good correlation with the back exposures and were on average 7 times lower. Exposures on gonads and forehead were even lower, hardly measurable ranging from 0.00 mSv to 0.01 mSv. The exposures measured with TL and RPL dosimeters were in agreement within measurement uncertainty. The relative large standard deviation has been caused by a series of factors (the irradiation conditions, patients' physical characteristics, work methods etc.) which cannot be influenced upon in everyday work. To evaluate the real exposure of children to radiation, it is necessary to have statistically relevant number of measurements which was 20 in this work.

One factor influencing the exposure of patients i.e. the working method of the radiology technicians was studied (Knežević et al. 2011). For this aim the dose was measured on the back of 60 patients during PA projection of thorax carried out by 3 technicians (A, B, C) with TL and RPL dosimeters. The results indicated that 20 patients examined by technician B received higher doses (average 0.34 mSv) than the others examined by technicians A and C (average 0.14-0.17 mSv) (see full signs for RPL dosimeters in Figure 4). After consultation and education the measurement cycle was repeated on 40 new patients (20 and 10 for technician B, for technicians A and C, respectively). As shown in Figure 4 (clear signs) the doses on children patients for all technicians were reduced (average 0.14 mSv). The reduction was achieved by changing the tube voltage (93-100 kV compared to earlier 89-96 kV) and quantity of charge (2.1-2.5 mAs compared to 5.0-9.0 mAs) by technician B. These results indicate the importance of permanent dose measurement on patients as well as the education of the personal.

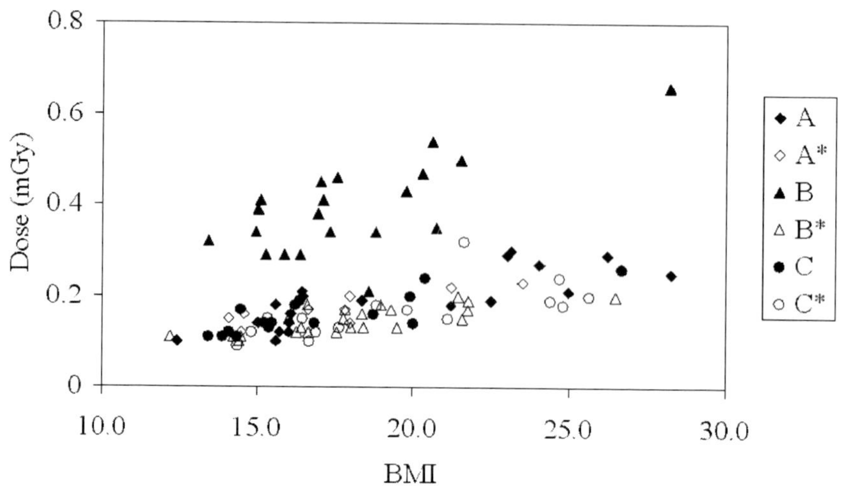

Figure 4. Doses measured by radiophotoluminescent dosimeter (RPL) on back of children patients during thorax diagnostics as a function of the body mass index (BMI). A, B, C are the symbols for the 3 technician. The results obtained after changing the protocol are marked with symbol *.

3.2. Biodosimtery Data using Cytogenetic Techniques after Patient's Exposure in X- Ray Diagnostic of Thorax

This chapter also illustrates the results of the alkaline comet assay and the CBMN assay as measure of genomic instability on human peripheral blood lymphocytes obtained from children exposed to diagnostic X-ray examination. Results for the parameters of the comet assay and the CBMNs assay before and after the radiation are presented in Figure 5. Basic statistics revealed that after exposure to diagnostic X-rays all of the tested parameters (tail length, tail intensity, tail moment, number of micronuclei, nucleoplasmic bridges and nuclear buds) were significantly higher in compare to measured values prior to X-ray examination.

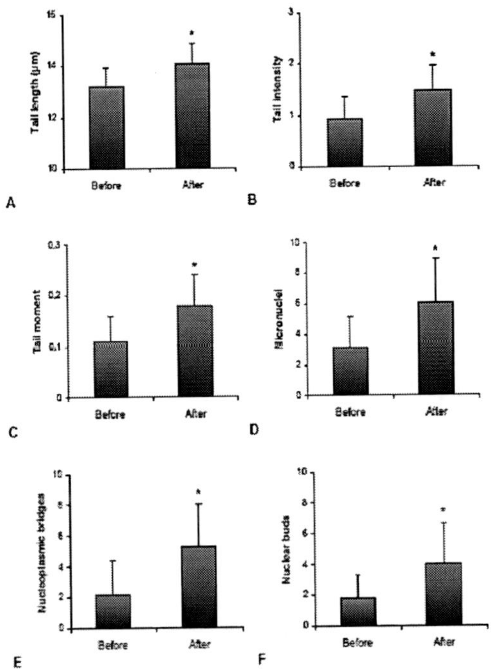

Figure 5. Mean group (±SD) values for alkaline comet assay parameters (tail length, tail intensity and tail moment), and cytokinesis-block micronucleus assay parameters (micronuclei, nucleoplasmic bridges and nuclear buds) in peripheral blood lymphocytes of children with pulmonary diseases before and after exposures to X-ray diagnostic examination. * Statistically significant compared to before X-ray diagnostics examination (non-parametric Mann-Whitney U test, $P<0.05$).

Since ionizing radiation plays a significant role in medical diagnostics, and is frequently needed for the diagnosis of diseases in pediatric patients, good medical practice dictates that children must be protected from unnecessary radiation. In that manner, this chapter is dealing with application of different dosimetry systems in order to evaluate their sensitivity and applicability in medical practice. Additionally, the impact on genomic instability prior to and after the diagnostic procedure was evaluated with the standard biomarkers for DNA and chromosomal alterations such as the alkaline comet assay and the CBMN assay.

The benefits of each and every single exam have to be clear when indicating one. The most important thing to have in mind is the risk–benefit ratio or simplified is the procedure going to give an answer to the pathology suspected. If the answer to that question is positive then the exam should be performed, if the answer is negative then a different diagnostic procedure should be taken in account. Often in medical reality the facts are not so simple and the conditions have to be diagnosed with combined exams and procedures or followed up frequently depending on the severity of each disease. All diagnostic procedures give answer to specific pathology and are of tremendous help in everyday life of a doctor (UNSCEAR 2000; Fučić et al. 2008; Hall & Brenner 2008; Gilbert 2009).

The radiation and dose are quantitated in many ways, such as entrance surface dose, air kerma, dose area product, dose length product etc. Organ specific doses are estimated by using phantoms with internal dosimeters or Monte Carlo software respectively. The doses that estimate direct harm from radiation for the patients are expressed as effective radiation dose. Effective radiation dose can give us a ground to estimate potential cancer risk or hereditary effects of procedures. For better understanding these are often compared the background radiation dose we are exposed to on a daily base. Doses vary in their range, not only for different procedures, but for one type of examination as well. No patient is the same and radiology technicians performing the exam do not have the same level of education or expertise and furthermore machines have to be maintained properly (UNSCEAR 2000; Hall & Brenner 2008; Stecker et al. 2009).

The lowest doses are in plain radiography where the effective doses vary between 0.001 mSv for a distal extremity image and 1.5 mSv for an image of lumbar spine. Using fluoroscopy to obtain images and view the anatomy in real time a higher dose is obtained by the patient ranging between 0.7 mSv and 18 mSv depending on the region examined and the pathology itself as well as the expertise of the doctor or radiology technician performing the exam. CT effective doses also depend on many different already mentioned factors and range between 1-2 mSv for a head scan and sometimes up to 30 or even 40 mSv for studies of pulmonary embolism and specific abdominal studies where the process has to be repeated with and without contrast. Effective doses are highest during interventional procedures and mostly depend on the complexity of the pathology and therapeutical intervention itself. The doses here can range to 180 mSv (average 70 mSv) for intrahepatic portosystemic shunt placement (Brugmans et al. 2002; Brix et al. 2003; Coles at al. 2006; Hall & Brenner 2008; Mettler et al. 2008; Seidenbusch & Schneider 2008; Gilbert 2009; Stecker et al. 2009).

As we can see there are many things the indicating doctor has to take in account when ordering one of the diagnostic procedures that use radiation. The ordering physician has to be aware of the risk-benefit ratio for every single patient with his own, unique pathology or condition. The principles on the other hand seem simple as the acronym ALARA indicates, which means As Low As Reasonably Achievable. In the recent 20 years the indications and ordered exams have multiplied in the world and today it is a question of not only properly executed exams to keep the doses low, but more of having the right indication for the exam. In children radiology we encounter more problems. The first one is of course the higher biosensitivity of children and the lower compliance. Also as children have a longer life expectancy the cumulative effects of radiation have to be considered and radiation has to be avoided as long as it is reasonable (Willis & Slovis 2004).

Pediatric conventional radiography is used to obtain images of mostly bones of the body and the intrathoracal or intraabdominal pathology. The most frequent examination of those is the standard summation image of intrathoracal organs, which gives a radiologist a view primarily into lungs and heart pathology and secondarily ribcage, thymus pathology. This examination is mostly performed because of different lung diseases of which infections are the most important. The effective dose of such examination is low and therefore has a low risk of causing cancer or hereditary issues. This on the other hand does not mean that such an examination is harmless and should be performed whenever. As every single child is different so are its genetic repair and resistance mechanisms and we can not foresee how the cumulative effect will potentially harm. That is why in pediatric plain radiography the standards have to be taken absolutely serious and every procedure has to be performed by the best known standard. This naturally includes the right indication, the best knowledge of the radiology technicians performing the examination and the best possible equipment firmly holding on the ALARA principle (Cook et al. 2001; Willis & Slovis 2004; Alt et al. 2006; Fučić et al. 2008; Hall & Brenner 2008; Seidenbusch & Schneider 2008; Gilbert 2009).

CONCLUSION

In conclusion, doses measured with TL and RPL dosimeters showed satisfactory agreement and both dosimetry methods are suitable for dosimetric measurements in X-ray diagnostics. The main advantage of TLD is the nearly tissue equivalence. The advantages of radiophotoluminescent dosimeters include negligible fading, high uniformity and possibility of repeating a dosimeter reading if necessary. The results indicate the importance of permanent dose measurement on patients as well as the education of the personal in order to improve radiation protection of children. Additionally, results gained by the comet assay and micronucleus test in children after diagnostic examination revealed that even the low dose X-rays can induce significant alterations in DNA molecule as well as on the chromosome level, since all the tested parameters were significantly increased after the radiation. Both the comet assay and the micronucleus test have proved to be valuable techniques of biological dosimetry and could be useful in acute events of pediatric genotoxic exposures. In addition, the simplicity, rapidity, and sensitivity of these methods could be used for purpose of monitoring the genetic damage in children often exposed to diagnostics or therapeutic procedures. Finally it is important to remember that the best possible protection from unnecessary exposure to radiation for patients is a right exam indication.

ACKNOWLEDGMENTS

Authors wish to thank all the volunteers that participated in this study and are grateful to Chiyoda Technol Corporation, Japan for providing the RPL dosimetry equipment. The study is a part of a research project supported by the Ministry of Science, Education and Sports of the Republic of Croatia (Grant no. 022-0222148-2125, 277-27770966-0962, 098-0982904-2954).

REFERENCES

Akahane, K; Hashimoto, M; Hada, M; Takano, Y; Ono, K; Kai, M; Kusama, T. Radiation doses to neonates during X ray computed tomography examinations. *Radiat Prot Dosimetry,* 2003, 103, 41-45.

Als-Nielsen, J; McMorrow, D. *Elements of Modern X-ray Physics*. Chicester, UK: John Wiley & Sons Ltd.; 2001.

Alt, CD; Engelmann, D; Schenk, JP; Troeger, J. Quality control of thoracic X-rays in children in diagnostic centers with and without pediatric-radiologic competence. *Rofo*, 2006, 178, 191-199.

Araki, F; Ikegami, T; Ishidoya, T; Kubo, HD. Measurements of gamma-knife helmet output factors using a radiophotoluminescent glass rod dosimeter and a diode detector. *Med Phys*, 2003, 30, 1976-1981.

Araki, F; Moribe, N; Shimonobou, T; Yamashita, Y. Dosimetric properties of radiophotoluminescent glass rod detector in high-energy photon beams from a linear accelerator and cyber-knife. *Med Phys*, 2004, 31, 1980-1986.

Brix, G; Nagel, HD; Stamm, G; Veit, R; Lechel, U; Griebel, J; Galanski, M. Radiation exposure in multi-slice versus single-slice spiral CT: results of a nationwide survey. *Eur Radiol*, 2003, 13, 1979-1991.

Brugmans, MJP; Buijs, WC; Geleijns, J; Lembrechts, J. Population exposure to diagnostic use of ionizing radiation in the Netherlands. *Health Phys*, 2002, 82, 500-509.

Burgkhardt, B; Röber, HG; Piesch, E. Phosphate glass energy compensation filters for the measurement of operational dose quantities. *Radiat Prot Dosimetry*, 1983, 6, 287-289.

Burgkhardt, B; Ugi, S; Vilgis, M; Piesch, E. Experience with phosphate glass dosimeters in personal and area monitoring. *Radiat Prot Dosimetry*, 1996a, 66, 83-88.

Burgkhardt, B; Festag, JG; Piesch, E; Ugi, S. New aspects of environmental monitoring using flat phosphate glass and thermoluminescence dosemeters. *Radiat Prot Dosimetry*, 1996b, 66, 187-192.

Coles, DR; Smail, MA; Negus, IS; Wilde, P; Oberhoff, M; Karsch, KR; Baumbach, A. Comparison of radiation doses from multislice computed tomography coronary angiography and conventional diagnostic angiography. *J Am Coll Cardiol*, 2006, 47, 1840-1845.

Collins, AR; Dusinská, M; Franklin, M; Somorovská, M; Petrovská, H; Duthie, S; Fillion, L; Panayiotidis, M; Raslová, K; Vaughan, N. Comet Assay in human biomonitoring studies: reliability, validation, and applications. *Environ Mol Mutagen*, 1997, 30, 139-146.

Collins, AR. Comet assay for DNA damage and repair: principles, applications and limitations. *Mol Biotechnol*, 2004, 26, 249-261.

Collins, AR; Oscoz, AA; Brunborg, G; Gaivão, I; Giovannelli, L; Kruszewski, M; Smith, CC; Stetina, R. The comet assay: topical issues. *Mutagenesis,* 2008, 23, 143-151.

Cook, JV; Kyriou, JC; Pettet, A; Fitzgerald, MC; Shah, K; Pablot, SM. Key factors in the optimization of paediatric X-ray practice. *Br J Radiol*, 2001, 74, 1032-1040.

Daffner, RH. *Clinical radiology: The Essentials*. Baltimore, MD: Lippincott Williams & Wilkins; 2007.

Dhawan, A; Bajpayee, M; Parmar, D. Comet assay: a reliable tool for the assessment of DNA damage in different models. *Cell Biol Toxicol*, 2009, 25, 5-32.

Dusinska, M; Collins, AR. The comet assay in human biomonitoring: gene-environment interactions. *Mutagenesis*, 2008, 23, 191-205.

Einstein, AJ; Henzlova, MJ; Rajagopalan, S. Estimating risk of cancer associated with radiation exposure from 64-slice computed tomography coronary angiography. JAMA, 2007, 298, 317-323.

Fenech, M; Morley, AA. Measurement of micronuclei in lymphocytes. *Mutat Res*, 1985, 147, 29-36.

Fenech, M; Perepetskaya, G; Mikhalevich, L. A more comprehensive application of the micronucleus technique for biomonitoring of genetic damage rates in human populations - experiences from the Chernobyl catastrophe. *Environ Mol Mutagen.* 1997, 30, 112-118.

Fenech, M; Chang, WP; Kirsch-Volders, M; Holland, N; Bonassi, S; Zeiger, E. HUMN project: detailed description of the scoring criteria for the cytokinesis-block micronucleus assay using isolated human lymphocyte cultures. *Mutat Res*, 2003, 534, 65-75.

Fenech, M. Cytokinesis-block micronucleus assay evolves into a "cytome" assay of chromosomal instability, mitotic dysfunction and cell death. *Mutat Res*, 2006, 600, 58-66.

Fenech, M. Cytokinesis-block micronucleus cytome assay. *Nat Protoc,* 2007, 2, 1084-1104.

Fenech, M. The lymphocyte cytokinesis-block micronucleus cytome assay and its application in radiation biodosimetry. *Health. Phys*, 2010, 98, 234-243.

Fenech, M; Kirsch-Volders, M; Natarajan, AT; Surralles, J; Crott, JW; Parry, J; Norppa, H; Eastmond, DA; Tucker, JD; Thomas, P. Molecular mechanisms of micronucleus, nucleoplasmic bridge and nuclear bud formation in mammalian and human cells. *Mutagenesis*, 2011, 26, 125-132.

Fučić, A; Brunborg, G; Lasan, R; Jezek, D; Knudsen, LE; Merlo, DF. Genomic damage in children accidentally exposed to ionizing radiation: a review of the literature. *Mutat Res*, 2008, 658, 111-123.

Gajski, G; Ravlić, S; Capuder, Ž; Garaj-Vrhovac, V. Use of sensitive methods for detection of DNA damage on human lymphocytes exposed to p,p'-DDT: Comet assay and new criteria for scoring micronucleus test. *J Environ Sci Health B*, 2007, 42, 607-613.

Gajski, G; Garaj-Vrhovac, V; Oreščanin, V. Cytogenetic status and oxidative DNA-damage induced by atorvastatin in human peripheral blood lymphocytes: standard and Fpg-modified comet assay. *Toxicol Appl Pharmacol*, 2008, 231, 85-93.

Garaj-Vrhovac, V; Đurinec, M; Kopjar, N; Oreščanin, V. A survey on the cytogenetic status of the Croatian general population by use of the cytokinesis-block micronucleus assay. *Mutat Res*, 2008a, 649, 91-100.

Garaj-Vrhovac, V; Gajski, G; Ravlić, S. Efficacy of HUMN criteria for scoring the micronucleus assay in human lymphocytes exposed to a low concentration of p,p'-DDT. *Braz J Med Biol Res*, 2008b, 41, 473-476.

Garaj-Vrhovac, V; Gajski, G; Pažanin, S; Šarolić, A; Domijan, AM; Flajs, D; Peraica, M. Assessment of cytogenetic damage and oxidative stress in personnel occupationally exposed to the pulsed microwave radiation of marine radar equipment. *Int J Hyg Environ Health*, 2011, 214, 59-65.

Garaj-Vrhovac, V; Kopjar, N; Ražem, D; Vekić, B; Miljanić, S; Ranogajec-Komor, M. Application of the alkaline comet assay in biodosimetry: assessment of in vivo DNA damage in human peripheral leukocytes after a gamma radiation incident. *Radiat Prot Dosimetry*, 2002, 98, 407-416.

Gilbert ES. Ionising radiation and cancer risks: What have we learned from epidemiology?. *Int J Radiat Biol*, 2009, 85, 467-482.

Hall, EJ; Brenner, DJ. Cancer risks from diagnostic radiology. *Br J Radiol*, 2008, 81, 362-378.

Honnef, D; Wildberger, JE; Stargardt, A; Hohl, C; Barker, M; Günther, RW; Staatz, G. Multislice spiral CT (MSCT) in pediatric radiology: dose reduction for chest and abdomen examinations. *Rofo,* 2004, 176, 1021-1030.

Horowitz, YS. *Thermoluminescence and Thermoluminescence Dosimetry.* Boca Raton, FL: CRC Press; 1984.

Hsu SM; Yeh, SH; Lin MS; Chen WL. Comparison on characteristics of radiophotoluminescent glass dosemeters and thermoluminescent dosemeters. *Radiat Prot Dosimetry,* 2006, Vol 119, 327-331.

IAEA, International Atomic Energy Agency. *Assessment of occupational exposure due to external sources of radiation*. Safety guide. IAEA safety standards series No. RS-G-1.3. Vienna, Austria: IAEA; 1999.

IAEA, International Atomic Energy Agency. *Dosimetry in diagnostic radiology: An international code of practice*. Technical report series No. 457. Vienna, Austria: IAEA; 2007.

ICRP, International Commision on Radiological Protection. *ICRP Publication 60: 1990 Recommendations of the International Commission on Radiological Protection*. Annals oft he ICRP, Vol 21(1-3). Oxford, UK: Pergamon Press; 1991.

Knežević, Ž; Ranogajec-Komor, M; Miljanić, S. Effect of dopants on TL characteristics of LiF:Mg, Cu, P detectors. *Radiat Meas,* 2010, 45, 573-575.

Knežević. Ž; Beck, N; Milković, Đ; Miljanić, S; Ranogajec-Komor, M. Characterization of RPL and TL dosimetry systems and comparison in medical dosimetry applications. *Radiat Meas*, 2011 (in press).

Kopjar, N; Kašuba, V; Milić M; Rozgaj, R; Želježić, D; Gajski G, Mladinić, M; Garaj-Vrhovac, V. Normal and cut-off values of the cytokinesis-block micronucleus assay on peripheral blood lymphocytes in the Croatian general population. *Arh Hig Rada Toksikol*, 2010, 61, 219-234.

Lee, JH; Lin, MS; Hsu, SM; Chen, IJ; Chen, WL; Wang, CF. 2009 Dosimetry characteristics and performance comparisons: Environmental radiophotoluminescent glass dosemeters versus thermoluminescent dosemeters. *Radiat Meas*, 2009, 44, 86-91.

Lee, JI; Yang, JS; Kim, JL; Pradhan, AS; Lee, JD; Chung, KS; Choe, HS. Dosimetric characteristics of LiF:Mg,Cu,Si thermoluminescent materials. *Appl Phys Lett,* 2006, 89, 094110- 094113.

Maluf, SW. Monitoring DNA damage following radiation exposure using cytokinesis-block micronucleus method and alkaline single-cell gel electrophoresis. *Clin Chim Acta*, 2004, 347, 15-24.

McArt, DG; McKerr, G; Howard, CV; Saetzler, K; Wasson, GR. Modelling the comet assay. *Biochem Soc Trans*, 2009, 37, 914-917.

McKeever, SWS; Moscovitch, M; Townsend, PD. *Thermoluminescence Dosimetry Materials - Property and Uses*. Ashford, UK: Nuclear Technology Publishing; 1994.

Mettler, FAJr; Huda, W; Yoshizumi, TT; Mahesh, M. Effective doses in radiology and diagnostic nuclear medicine: A catalog. Radiology, 2008, 248, 254-263.

Milković, Đ. Exposure of patients, possibilities of radiation protection and risk estimation in X-ray diagnostics of respiratory tract of children. Ph.D. Thesis, Medical Faculty, University of Zagreb; 1998.

Milković, Đ; Ranogajec-Komor, M; Krstić-Burić, M; Hebrang A. Mit Thermolumineszenz-Dosimetern gemessene Hautdosen bei Thorax-Rontgenaufnahmen von Kindern und Jugendlichen. *Atemw Lungenkrkh*, 1991, 17, 67-72.

Milković, Đ; Garaj-Vrhovac, V; Ranogajec-Komor, M; Miljanić, S; Gajski, G; Knežević, Ž; Beck, N. Primary DNA damage assessed with the comet assay and comparison to the absorbed dose of diagnostic X-rays in children. *Int J Toxicol,* 2009, 28, 405-416.

Miljanić, S; Ranogajec-Komor, M; Knežević, Ž; Vekić, B. Main dosimetric characteristics of some "tissue-equivalent" TL detectors. *Radiat Prot Dosimetry*, 2002, 100, 437-442.

Miljanić, S; Ranogajec-Komor, M; Blagus, S; Pálfalvi, J; Pázmándi, T; Deme, S; Szántó, P. Response of radiophotoluminescent dosimeters to neutrons. *Radiat Meas,* 2008, 43 1068-1071.

Møller, P. The alkaline comet assay: towards validation in biomonitoring of DNA damaging exposures. *Basic Clin Pharmacol Toxicol*, 2006, 98, 336-345.

Nose, T; Koizumi, M; Yoshida, K; Nishiyama, K; Sasaki, J; Ohnishi, T, Peiffert, D. *In vivo* dosimetry of high-dose rate brachytherapy: study on 61 head-and-neck cancer patients using radiophotoluminescence glass dosimeter. *Int J Radiat Oncol Biol Phys,* 2005, 61, 945-953.

Ohuchi, H; Satoh, T; Eguchi, Y; Mori, K. Preliminary study of using imaging plates to map skin dose of patients in interventional radiology procedures. *Radiat Prot Dosimetry*, 2005, 117, 432-439.

Olive, PL. Impact of the comet assay in radiobiology. *Mutat Res*, 2009, 681, 13-23.

Ono, K; Yoshitake, T; Akahane, K; Yamada. Y; Maeda, T; Kai, M; Kusama, T. Comparison of a digital flat-panel versus screen-film, photofluorography and storage-phosphor systems by detection of simulated lung adenocarcinoma lesions using hard copy images. *Br J Radiol*, 2005, 78, 922-927.

Perks, J; Gao, M; Smith, V; Skubic, S; Goetsch, S. Glass rod detectors for small field, stereotactic radiosurgery dosimetric audit. *Med Phys*, 2005, 32, 726-732.

Piesch, E. Development in Radiophotoluminescence Dosimetry. In: Attix, FH, editor. *Topics in Radiation Dosimetry*. New York, NY: Academic Press: 1972; 461-532.

Piesch, E; Burghardt, B; Vilgis, M. Photoluminescence dosimetry: progress and present state of art. *Radiat Prot Dosimetry*, 1990, 33, 215-226.

Piesch, E; Burgkhardt, B. Photoluminescence dosimetry: the alternative in personnel monitoring. *Radioprotection,* 1994, 29, 39-67.

Piesch, E; Burghardt, B; Vilgis, M. Performance of the new photoluminescent glass dosimetry system at the IAEA intercomparison 1990/91 for individual monitoring. *Radiat Prot Dosimetry,* 1994, 54, 109-117.

Ranogajec-Komor, M. Thermoluminescence Personal and Medical Dosimetry. In: Zaidi, MK; Mustafaev, I, editors. *Radiation Safety Problems in the Caspian Region*. Dordrecht, The Netherlands: Kuwler Academic Publishers; 2004, 177-190.

Ranogajec-Komor, M; Knežević, Ž; Miljanić, S; Vekić, B. Characterisation of radiophotoluminescent dosimeters for environmental monitoring. *Radiat Meas*, 2008, 43, 392-396.

Ranogajec-Komor, M; Muhiy-Ed-Din, F; Milković, D; Vekić, B. Thermoluminescence characteristics of various detectors for X ray diagnostic measurements. *Radiat Prot Dosimetry*, 1993, 47, 529-534.

Ranogajec-Komor, M; Nikodemova, D; Horvathova, M; Knežević, Ž; Milković, Đ. Radiation exposure of children during radiodiagnostic examination of chest. In: Sabol, J, editor. *Proceedings of the IRPA Regional Symposium on Radiation in Neighbouring Countries of Central Europe*. Prague, Czech Republic: Czech Technical University; 1997, 298-300.

Ranogajec-Komor M; Osvay M. Dosimetric Characteristics of different TL phosphors. *Radiat Prot Dosimetry*, 1986, 17, 379-384.

Röntgen, WC. Über eine neue Art von Strahlen. *Annalen der Physik*, 1898, 300, 1-11.

Sacks, D; McClenny, TE; Cardella, JF; Lewis, CA. Society of Interventional Radiology clinical practice guidelines. *J Vasc Interv Radiol.* 2003, 14, 199-202.

Schulman, JH; Etzel, HW. Small-volume dosimeter for X-rays and gamma rays. *Science* 1953, 118, 184-186.

Seidenbusch, MC; Schneider, K. Radiation exposure of children in pediatric radiology. Part 4: Entrance doses achieved during the X-ray examination of the chest. *Rofo*, 2008, 180, 1082-1103.

Simpson, AK; Whang, PG; Jonisch, A; Haims, A; Grauer, JN. The radiation exposure associated with cervical and lumbar spine radiographs. *J Spinal Disord Tech*, 2008, 21, 409-412.

Singh, NP; McCoy, MT; Tice, RR; Schneider, EL. A simple technique for quantitation of low levels of DNA damage in individual cells. *Exp Cell Res*, 1988, 175, 184-191.

Sirota, NP; Kuznetsova, EA. The comet assay application in radiobiological investigations. *Radiats Biol Radioecol*, 2010, 50, 329-339.

Slovis, TL. The ALARA concept in pediatric CT: Myth or reality? *Radiology,* 2002, 223, 5-6.

Stecker, MS; Balter, S; Towbin, RB; Miller, DL; Vano, E; Bartal, G; Angle, JF; Chao, CP; Cohen, AM; Dixon, RG; Gross, K; Hartnell, GG; Schueler, B; Statler, JD; de Baère, T; Cardella, JF; SIR Safety and Health Committee; CIRSE Standards of Practice Committee. Guidelines for patient radiation dose management. *J Vasc Interv Radiol*, 2009, 20, 263-273.

Taguchi, K; Anno, H. High temporal resolution for multislice helical computed tomography. *Med Phys,* 2000, 27, 861-872.

TGC, Asahi Techno Glass Corporation. *Glass Dosimetry System FGD-200 Series Instruction Manual* (Ver.2). Shizuoka: Asahi Techno Glass Corporation; 2003.

TGC, Asahi Techno Glass Corporation. *RPL Glass Dosimeter - Small Element System.* Shizuoka: Asahi Techno Glass Corporation; 2004.

Thomas, P; Umegaki, K; Fenech, M. Nucleoplasmic bridges are a sensitive measure of chromosome rearrangement in the cytokinesis-block micronucleus assay. *Mutagenesis*, 2003, 18, 187-194.

Thomsen, HS; Webb, JAW. *Contrast Media: Safety Issues and ESUR Guidelines.* Berlin Heidelberg: Springer Verlag; 2009.

Tice, RR; Strauss, GH. The single cell gel electrophoresis/comet assay: a potential tool for detecting radiation-induced DNA damage in humans. *Stem Cells*, 1995, 13, 207-214.

Tsuda, M. A few remarks on photoluminescence dosimetry with high energy X rays. *Igaku Butduri*, 2000, 20, 131-139.

Umegaki, K; Fenech, M. Cytokinesis-block micronucleus assay in WIL2-NS cells: a sensitive system to detect chromosomal damage induced by reactive oxygen species and activated human neutrophils. *Mutagenesis*, 2000, 15, 261-269.

UNSCEAR, United Nations Scientific Committee on the Effects of Atomic Radiation. *Sources and Effects of Ionizing Radiation*. UNSCEAR 2000 Report II.-Effects. Vienna, Austria: 2000.

Valverde, M; Rojas, E. Environmental and occupational biomonitoring using the Comet assay. *Mutat Res*, 2009, 681, 93-109.

Varčenja, VZ; Gubatova, DJ; Lubocky, II; Plume, AH. Kriterien der Strahlenrisikos bei Röntgenuntersuchungen von Kindern. *Radiol Diagn*, 1988, 29, 747-752.

Vral, A; Fenech, M; Thierens, H. The micronucleus assay as a biological dosimeter of in vivo ionising radiation exposure. *Mutagenesis*, 2011, 26, 11-17.

Williams, JR. The interdependence of staff and patient doses in interventional radiology. *Br J Radiol*, 1997, 70, 498-503.

Willis, CE; Slovis, TL. The ALARA concept in pediatric CR and DR: dose reduction in pediatric radiographic exams – A white paper conference executive summary; *Pediatr Radiol*, 2004, 34, 162-164.

Yokota, R; Muto, Y. Simplified method of finding quantum radiation energy and its incident direction using a single dosimeter glass. *Health Phys*, 1971, 21, 597-598.

Zhuo, W; Liu, W; Huang, G; Zhu, G; Ma, G. Comparisons of dosimetric properties between GD-300 series of radiophotoluminescent glass detectors and GR-200 series of thermoluminescent detectors. *Nucl Sci Tech*, 2007, 18, 362-365.

Zoetelief, J; Julius, HW; Christensen, P. *Recommendations for patient dosimetry in diagnostic radiology using TLD*. Rep. EUR 19604, European Commission. (IAEA-CN-96-46P). Luxembourg: 2000.

In: Radiation Exposure
Editors: D. Balenovic and E. Stimac
ISBN 978-1-62100-869-9

Chapter 3

HEMOPOIESIS IN RESIDENTS OF THE TECHA RIVERSIDE VILLAGES AFTER LONG-TERM LOW-DOSE RATE RADIATION EXPOSURE

Alexander Akleyev,* Tatyana Varfolomeyeva, Ludmila Krestinina, and Georgiy Dimov

Urals Research Center for Radiation Medicine,
Chelyabinsk, Russian Federation

ABSTRACT

This Chapter focuses on the analysis of data resulting from almost 60 years of studies aimed at assessment of the hemopoiesis status in Techa riverside residents exposed to low-dose rate (LDR) radiation in the range from low to high doses. The red bone marrow (RBM) doses (individual values: 0.01-6.69 Gy; mean value - 0.89 Gy) were due to intakes of ^{90}Sr in drinking water and local food products, and also to external γ-exposure. The status of hemopoiesis (cellular composition of the RBM and peripheral blood, incidence of leukemia and RBM hypoplasia) was monitored from 1951 through 2008. Peripheral blood cell counts performed for exposed riverside residents during the period of maximum radiation exposures (1951-1953) displayed a significant (p<0.0001)

* Address for correspondence: Prof. Alexander Akleyev, M.D., URCRM, 68-A Vorovsky St., Chelyabinsk 454076 Russia. TEL: +7-351-232-7914, FAX: +7-351-232-7913, E-mail: akleyev@urcrm.ru.

decrease in the proportion of leukocytes (neutrophils and lymphocytes) and thrombocytes. Peripheral blood erythrocyte count was maintained at the normal level, owing to the efficacy of compensatory processes in the bone marrow, including increased rates of erythrokaryocyte proliferation and maturation. The development of peripheral blood granulocytopenia resulted from a delay in the differentiation of neutrophilic bone-marrow granulocytes, a marked increase in the frequency of lethal abnormalities in bone-marrow neutrophils, and pathological mitoses. The periods of recovery of the blood cell composition differed considerably between different blood cell series, and depended on exposure dose rates, degree of the primary inhibition of hemopoiesis, as well as on individual physiological characteristics of the exposed persons (gender, age at initial exposure). At late time, an increased risk of leukemia was registered.

Keywords: Radiation exposure; hemopoiesis; bone marrow; recovery; cytopenia; leucopenia

INTRODUCTION

It is well known that, owing to the high radiosensitivity of the hemopoietic system, its inhibition is a critical phenomenon determining the severity of the damage induced by high doses of radiation. So far, the effects of acute external γ-exposure have been studied most thoroughly: initial reactions of hemopoietic cells and the human hemopoietic system as a whole, mechanisms and dynamics of recovery processes going on in the hemopoietic system, as well as the radiogenic risk of leukemias as a most likely long-term effect [1, 2]. Far less is known about the reactions of hemopoiesis to chronic (for multiple years) low dose-rate exposures in man. Experimental studies in beagles have provided evidence of high plasticity of the hemopoietic system and its sufficient capacity to adapt to protracted irradiation, even in lifelong exposures [3, 4].

The purpose of the study was to trace the dynamics of hemopoiesis and late effects for over 60 years after the onset of LDR radiation exposure in man. Radiation exposure of Techa riverside residents was associated with the operation of the Mayak facility which is located in the center of Russia, in Chelyabinsk Oblast. Because of the lack of reliable technologies for reprocessing and storing radioactive waste, beginning in March 1949 liquid radioactive wastes from the Mayak facility were discharged into the Techa River. From 1949 to 1956 the Mayak combine discharged about 114 PBq of liquid radioactive waste into the Techa River. The major releases occurred over the period of 1950-1951 [5]. The long-lived ^{90}Sr and ^{137}Cs were the main contributors to the environmental contamination. The releases of radioactive waste resulted in a heavy contamination of the river system (water, bottom sediments, and flood plain).

Residents of the riverside settlements (about 30 thousand people) were chronically exposed to LDR radiation, both external, due to γ-radiation, and internal due to intakes of radionuclides in drinking water and local foodstuffs. Critical, in terms of exposure doses received by residents of the Techa riverside villages, was the hemopoietic system. Maximum exposure doses to red bone marrow (RBM) reached 9.0 Gy which is about 50 times higher as compared to the average worldwide lifelong dose from natural radiation sources. Medical follow-up of the exposed population which has been conducted since 1951 by the physicians of the clinic of the Urals Research Center for Radiation Medicine (URCRM) includes complete blood studies as an obligatory test, and, in a number of cases, RBM studies (by medical indications) [6].

The Techa River cohort (TRC) was established in 1967 by the URCRM with the purpose to study radiation-induced health effects of chronic environmental radiation exposure [7]. The database maintained at the URCRM has served as a key information source for investigating the state of hemopoiesis for residents of contaminated areas for about 60 years of the follow-up.

CHARACTERISTICS OF THE TECHA RIVER COHORT AND THE FOLLOW-UP

The population of the Techa riverside villages was exposed to external and internal radiation. External radiation exposure from gamma-emitting radionuclides (^{137}Cs, ^{95}Zr, ^{95}Nb, ^{103}Ru, ^{106}Ru, and other) was a consequence of contamination of the river water and floodplains. Internal exposure resulted from consumption of water, milk and other food products contaminated with radionuclides (primarily ^{89}Sr, ^{90}Sr, ^{137}Cs). Internal exposure was primarily a result of ^{90}Sr incorporation into the bone structure, for which the bone marrow was the main target-organ. On the average, 92% of the total red bone marrow dose was due to internal exposure. About 80% of the total red bone marrow dose (RBM) was accumulated over the first decade of exposure.

The TRC includes 29,756 persons, born before 1.1.1950, who lived in any of the 41 Techa riverside villages during the period from 1950 to 1960. More than 80% of the cohort members lived near the river between 1950 and 1953, the period of the highest radiation exposure. The cohort includes individuals of all ages, is about 58% female and is comprised of two major ethnic groups, with 80% being of Slav origin, and 20% of Tartar or Bashkir origin. About 40% of the cohort members were aged under 20 years when first exposed.

Table 1. Vital status for TRC members as of 31.12.2005

Vital Status	N	%
Alive	6,363	21.4
Deceased	16,619	55.9
Cause of death known	*15,025*	*90.4*
Vital status unknown	2,219	7.5
Distal migrants	4,555	15.3
Total	29, 756	100.0

Table 1 summarizes vital status and follow-up data for TRC members through 31.12.2005. At that time, 21% of the cohort members were alive and living in the catchment area. Among the 56% of cohort members identified as deceased, the cause of death was established for 90%. Vital status at the end of follow-up was unknown for only 7.5% of cohort members remaining in the study catchment area, but the follow-up was censored for about 15% of the cohort when they emigrated from the catchment area (referred to as distal migrants).

The state of blood and RBM cell composition in exposed persons was studied over the period from 1951 through 2008. The study subcohort comprising 4,629 TRC members received regular examinations to evaluate the cellular composition of peripheral blood (Table 2).

Patients diagnosed with diseases that could affect hemopoiesis parameters (tumors and hematological conditions) were not included in the study of the dynamics of blood cell composition. Also, excluded from the analyses were data on the blood parameters measured in patients during the period of aggravations or exacerbations of their chronic inflammatory diseases.

Table 2. Basic characteristics of the groups compared

Group	Ethnic composition				Gender				Mean age as of 1950, years	Mean age as of 2008, years	Cumulative dose to RBM as of 2005, Gy
	Turkic		Slavs		Male		Female				
	Per-sons	%	Per-sons	%	Per-sons	%	Per-sons	%			
Exposed n=4,629	2,016	43.6	2,613	56.4	1,957	42.3	2,672	57.7	19.0±0.1 (1-67)	65.2±0.3 (52-88)	0.89±0.01 (0.01-6.69)
Control n=1,042	304	29.2	738	70.8	380	36.5	662	63.5	18.4±0.4 (1-64)	63,6±0,6 (52-85)	0

In brackets: range of individual values.

The study group included 1,957 exposed men (42.3%) and 2,672 exposed women (57.7%). As of the start of exposure, the age of the study group members ranged from 1 to 67 years with the mean of 19.0 years, and by the end of the follow-up (2008) they reached from 52 to 88 years with the mean of 65.2 years.

The control group was composed of people living in the same villages as members of the study group. Red bone marrow (RBM) doses received by the control individuals did not exceed 5 mGy/yr during the period of the maximum radiation exposure. The study and the control groups were comparable in terms of proportions of people representing different age categories and genders. The control group consisted of 1,042 persons, including 380 (36.5 %) men and 662 (63.5 %) women. As of 1950, the age of the control group members ranged from 1 to 64 years with the mean of 18.4, and by the end of the follow-up (2008), they reached 52-85 years with the mean of 63.6 years. It should be noted that the analysis was based on the hematological data resulting from medical examinations of persons aged 16 or older.

Cytological studies of red bone marrow were performed for 44 Techa riverside residents aged 16-48 years with the mean of 24.6 years. RBM biopsy specimens obtained in 1954-1956 with sternal puncture were reviewed retrospectively. Male patients accounted for 50% of the total examined subjects. The control group included 16 unexposed individuals aged 16-49 years with the mean of 24.8 years, and 5 of them underwent thorough medical examinations since they were bone-marrow donors, and examinations for presumed hematological conditions in the remaining 11 produced negative result

Throughout the follow-up period, hematological studies were performed using standard diagnostic laboratory methods [8, 9]. The status of the hemopoiesis was evaluated based on cytological studies of RBM aspirates (scoring 200 cells), and by estimation of cell ratio in different hemopoietic cell lineages [10] (scoring 1,000 cells per each cell lineage). The hematological examinations included identification of RBM cells at various phases of the erythroid series differentiation: erythroblast (E1) - proerythroblast (E2) – basophilic normocyte (E3) – polychromatophilic normocyte (E4) – oxyphilic normocyte (E5), neutrophylic series: myeloblasts (M1) – promyelocyte (M2) – maternal myelocyte (M3) daughter myelocyte (M4) - metamyelocyte (M5) – stab neutrophil (M6) – segmented neut- rophil (M7), as well as types of cell abnormalities. The hematological studies focused on RBM cell abnormalities (binuclear and giant cells, fragmentation of nuclei, karyolysis, pyknosis of nuclei, micronuclei and cytoplasmic bridges), defective mitoses (chromosome stickiness and interchromosomal bridges) and proliferative activity of the erythrokaryocytes and neutrophilic granulocytes. The studies also included evaluation of proliferative activity of erythrokaryocytes and neutrophilic granulocytes and estimation of the mitotic index (MI) of cells at the stages E1+E2, E3, E4 and M1+M2, M3+M4, and also the common proliferative activity of morphologically identifiable erythrokaryocytes and neutrophilic granulocytes.

Bone marrow doses were estimated on the basis of the Techa River dosimetry system 2009 (TRDS-2009) [5]. The system provides individualized internal and external organ dose estimates and information on the uncertainty of these estimates. The dynamics of RBM dose rate was characterized by maximum individual values reaching 2.03 Gy in 1951 with a subsequent dramatic decrease in 1956, and later on by a gradual decrease, after the releases had been terminated. RBM doses had mostly accumulated by 1960, and in the study group members the cumulative RBM doses had reached the maximum of 6.69 Gy (Figure 1). The dynamics of dose rates characterized by maximum values in the period of releases and a consequent long-term decrease to background values corresponds to a typical model of accidental exposures of the general population resulting from releases of long-lived ^{90}Sr and ^{137}Cs into the environment.

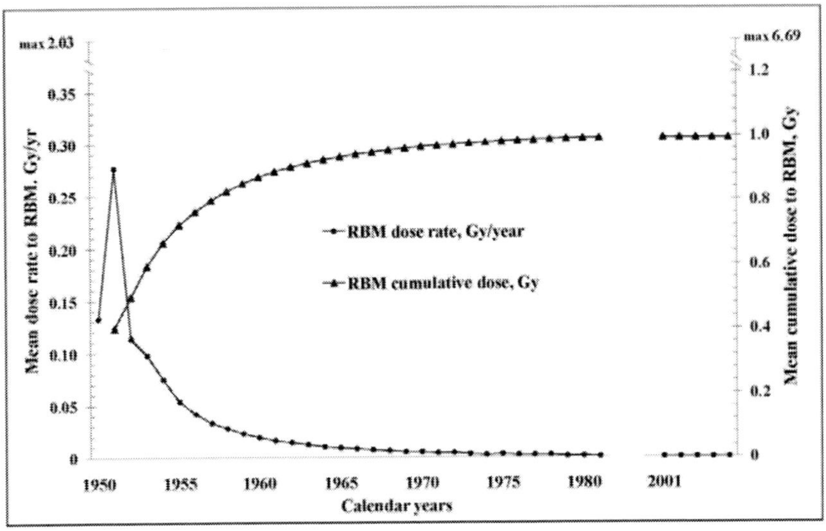

Figure 1. Dynamics of average cumulative dose and dose rate to RBM for residents of contaminated areas on the Techa River (TRDS -2009).

The incidence of late effects of RBM exposure (leukemia and RBM hypoplasia) among all TRC members resident in the catchment area (Chelyabinsk and Kurgan Oblasts) was assessed based on data resulting from several decades of the follow-up [11]. Cases of leukemia and RBM hypoplasia were ascertained from four main sources: URCRM medical records, regional oncology clinics, regional health centers and death certificates. The URCRM was a key source of information on cases of leukemia and RBM hypoplasia in the first two decades of follow-up, providing data on 39% of all leukemia cases. Since 1956, leukemia notification forms from the Chelyabinsk and Kurgan oblast oncology dispensaries have been linked to the TRC cohort roster and about 37% of all cases have been identified from this source. Since the dispensaries are the major cancer treatment facilities in the study catchment area, they have become the primary source for newly diagnosed cases. About 6% of the total leukemia cases were ascertained from the records of health centers in Chelyabinsk and Kurgan oblasts. The remaining leukemia cases (19 %) have been ascertained from death certificates stored at the government offices of Chelyabinsk and Kurgan oblasts. With these data sources, the ascertainment of leukemia cases is largely complete since 1953 for cohort members residing in Chelyabinsk Oblast or Kurgan Oblast. A more detailed description of methods for information collection on vital status and disease incidence among TRC members is presented in [12].

Cohort members were considered to be at risk from the earliest of January 1, 1953 or the date at which they first moved into one of the affected villages until their earliest of the date of leukemia diagnosis or death. Person-years were accumulated only during periods in which a person was known to be living in Chelyabinsk or Kurgan Oblasts. The radiation effect was described using excess relative risk models [13]. Linear, linear-quadratic and pure-quadratic dose-response models were considered. Age, birth cohort, gender, ethnicity, and oblast of initial exposure were evaluated as potential risk factors influencing leukemia baseline rates. Gender, age at entry, attained age, and ethnicity were considered as potential dose-effect modifiers. To allow estimation of radiogenic risk of leukemia, red bone marrow doses estimated based on the TRDS-2000 were used [14].

Comparison of respective continuous parameters between the exposed and control groups was performed using Student's *t*-test in case of normally distributed parameters, and Mann-Whitney *U*-test in other cases. Comparison of the frequency between the exposed and control groups was based on Pearson's χ^2 criterion and Fisher's exact test. The necessary calculations were performed using version 10 of the SPSS for Windows (SPSS Inc., USA).

Analyses of background rates and excess leukemia risk were carried out using Poisson regression methods and the Epicure software [13, 15] in a highly stratified table of cases and person-years for leukemia, with the exception of cases of chronic lymphatic leukemia (non-CLL), and a single case of CLL to provide the most stable site specific estimates feasible with this cohort at this time.

Bone Marrow Hemopoiesis in the Period of Maximum Exposure

It was established that RBM cellularity, relative proportion of neutrophil granulocytes and leuko-erythroblastic ratio observed in exposed persons and controls over the period 1954-1956 were comparable. The number of myelokaryocytes in the exposed group was in the range of 42.5-340.5 $\times 10^9$/L with the mean±SD of 132.0±72.7 $\times 10^9$/L, and that in the control group was in the range of 44-189 $\times 10^9$/L with the mean±SD of 102.3±42.3 $\times 10^9$/L. The relative proportion of RBM neutrophil granulocytes was 66.6±7.3% and 68.5±6.8% in the study and control groups, respectively. The proportion of RBM erythrokaryocytes was significantly increased (p=0.026) in the exposed group (24.1±7.6%) compared to the control group (20.5±5.4%). Leuko-erythroblastic ratio did not differ significantly between the study group (3.6) and the control group (4.2).

Conventional cytological analyses of RBM aspirates (Table 3) indicated a significant decrease in the number of erythroblasts, and a significant increase in oxyphilic normocytes in the exposed group compared to the control group. No significant difference was observed between the exposed and control groups in the myeloid cell series as well as in lymphocyte, monocyte and plasmocyte counts. Besides, there was no significant difference in the neutrophil maturation index (MI) between the exposed and control groups.

Normal cytology findings were observed during the period of maximum radiation exposure in 31.2 % of the control group, only in 2.3% of the exposed group members. Residents of the Techa riverside villages manifested a significantly (p=0.019) higher frequency (36.4%) of erythrocytic series hyperplasia (erythrokaryocytes > 26.5%) compared to that displayed by the controls (6.2%).

Studies of proliferative activity of RBM neutrophils and mitotic index of neutrophilic myelocytes did not reveal any difference between the exposed and control groups. The frequency (±SD) of mitoses per 1,000 neutrophilic RBM granulocytes was 3.4±2.5 and 3.7±1.2 in the exposed and control groups, respectively; the mitotic index (±SD) of myelocytes in the respective groups was 1.1±0.8% and 1.4±0.6%.

Table 3. Distribution of bone marrow cell composition in exposed and control groups in 1954-1956

Parameter	Exposed group	Control group
Undifferentiated blasts (%)	0.1±0.33[b]	0.03±0.12
Myeloblasts (%)	0.7±0.8	0.5±0.5
Promyelocytes (%)	1.2±1.0	0.8±0.5
Neutrophilic myelocytes (%)	18.0±4.4	17.1±5.1
Metamyelocytes (%)	15.9±3.7	17.5±4.6
Band neutrophils (%)	17.9±4.2	18.0±4.4
Segmented neutrophils (%)	12.9±5.1 *p=0.0528*[a]	15.8±4.9
Neutrophil maturation index	1.2±0.4	1.1±0.5
Total eosinophils (%)	3.7±2.2	3.1±1.9
Total monocytes (%)	0.07±0.17	0.2±0.4
Lymphocytes (%)	4.2±3.5 *p=0.0516*	6.5±3.7
Plasmocytes (%)	1.0±0.8	1.0±0.9
Erythroblasts (%)	0.08±0.2 *p=0.0055*	0.3±0.3
Pronormoblasts (%)	1.2±0.9	0.8±0.7
Basophilic normocytes (%)	3.6±2.0	3.3±2.1
Polychromatophlic normocytes (%)	8.5±4.5	8.3±4.0
Oxyphilic normocytes (%)	10.8±3.9 *p=0.0003*	6.8±3.5

[a]Based on Student's *t*-test comparing the exposed and control groups; [b]mean ± standard deviation.

The differentiation of neutrophilic RBM granulocytes in the two groups was assessed using the method of partial granulocytograms. During the period of maximum radiation exposures an increased production of myeloblasts (1.04% in exposed individuals *vs* 0.54% in controls, *p*=0.0051) and neutrophilic myelocytes (28.0% in exposed individuals *vs* 24.3% in controls,

p=0.0091) was observed (Figure 2). This fact indicates a delay in granulocyte maturation in individuals chronically exposed to radiation.

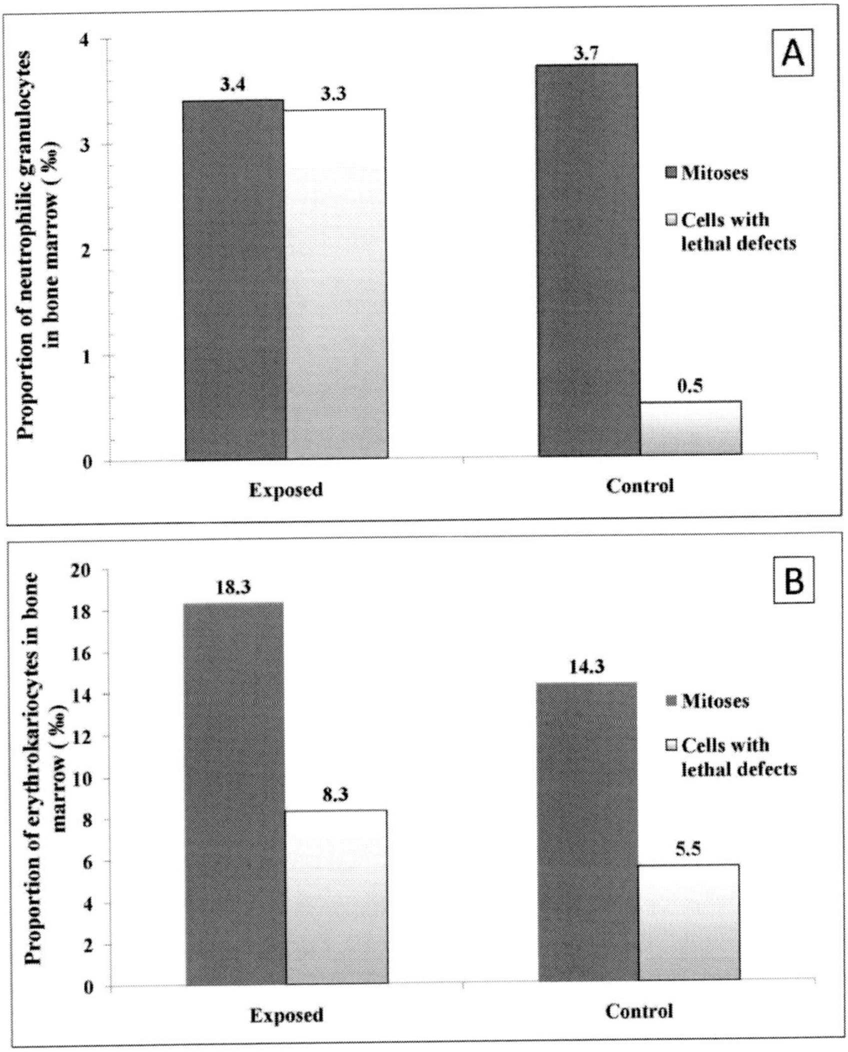

Figure 2. Frequency of mitoses and lethal abnormalities observed in chronically exposed subjects in the period of maximum radiation exposure (1951-1956): (A) in bone marrow neutrophilic granulocytes (binuclear cells, giant cells, fragmentation of nuclei, karyolysis); (B) in erythrokaryocytes (interchromosomal bridges, binuclear cells, micronuclei).

Proliferative activity of erythrokaryocytes was significantly higher in the exposed group than in the control group (Table 4). The mitotic indices (MIs) of basophilic normocytes and polychromatophilic normocytes were both higher in the exposed group than in the control group. Individuals with increased mitotic activity of erythrokaryocytes were seen significantly (p=0.0126) more frequently in the exposed group (31.8%) than in the control group (0%).

Table 4. Distribution of proliferative activity of erythrokaryocytes in exposed and control groups in 1954-1956

Parameter	Exposed group	Control group
Mitoses, per 1,000 erythrokcytes	18.3±5.7[b] *p=0.0028*[a] (7-33)[c]	14.3±3.4 (7-19)
MI (erythroblasts and proerythroblasts), %	8.3±4.9 (0-20)	7.2±4.9 (0-18.5)
MI (basophilic normocytes), %	9.6±5.1 *p=0.0013* (1.6-23.1)	6.1±2.9 (0-9.9)
MI (polychromatophilic normocytes), %	1.3±1.1 *p=0.0152* (0-4.2)	0.8±0.4 (0-1.7)

[a] Based on Student's t-test comparing the exposed and control groups; [b] mean ± standard deviation; [c] minimum-maximum.

It was established that in 1954-1956 the differentiation of erythrokaryocytes had undergone certain changes in the exposed group members (Figure 3). A statistically significant increase in oxyphilic normocytes (51.4% in the exposed group *vs* 42.4/% in the control group; p=0.0013) was observed along with a statistically significant decrease in polychromatophilic normocytes (33.6% in the exposed group *vs* 39.7% in the control group; p=0.017). In addition, a statistically significant reduction in erythroblasts was registered (0.6% in the exposed group *vs* 1.1% in the control group; p=0.0186).

Assessment of the erythropoiesis status made it possible to identify a significant number of chronically exposed persons (31.8%) in whom increased proliferation of erythroid cells was combined with increased production of mature erythrokaryocytes (cells at oxyphilic normocytes phase > 5.6%). Besides, 35.7% of them were noted to have developed hyperplasia of bone-marrow erythrocytic series (erythrokaryocytes in the bone marrow aspirate > 26.5%).

The frequency of pathological mitoses in myeloid and erythroid cell series (mostly chromosome stickiness) and lethal abnormalities in RBM hemopoietic cells registered in 1954-1956 was significantly higher in the exposed group than in the control group (Figure 4). Pathological mitoses were significantly more frequent in the exposed group than in the control group, especially in the erythroid cell series. Bridges between chromosomes were observed very infrequently and only in the erythroid cell series at annual dose-rate exceeding 0.1 Gy.

The results from the studies of abnormalities detected during the interphase in RBM neutrophils are presented in Table 5. Aberrant neutrophils were significantly more frequent in the exposed group than in the control group. Attention is drawn to a significantly increased number of cells with lethal aberrations (binuclear cells, giant cells, fragmentation of nuclei, karyolysis) in exposed group during the period of maximum radiation exposure. Compared to the control group, the exposed group showed more frequently such signs of apoptosis as pyknosis, fragmentation of nuclei and karyolysis in neutrophilic granulocytes of the bone marrow. Cells with karyolysis were observed only in RBM neutrophilic lineages in the exposed group.

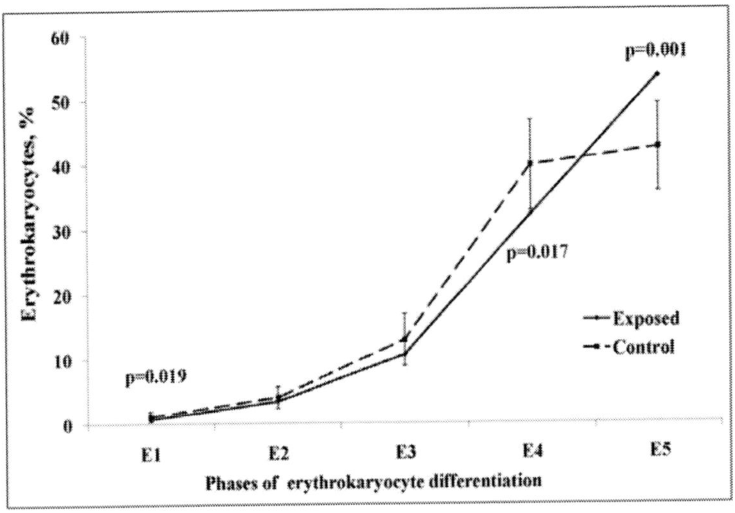

Figure 3. Distribution of bone marrow erythrokaryocytes by maturation phases registered in 1954-1956. The dashed lines indicate 1.0 SD.

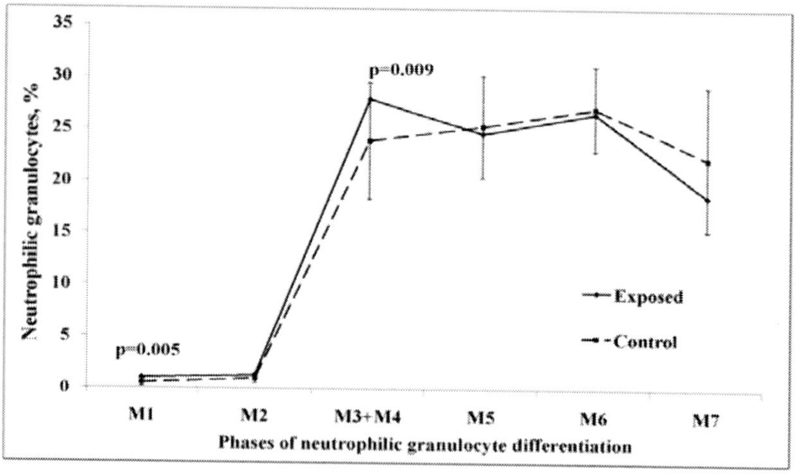

Figure 4. Distribution of bone marrow neuthrophilic granulocytes by maturation phases registered in 1954-1956. The dashed lines indicate 1.0 SD.

Table 5. Frequency of abnormalities per 1,000 bone marrow neutrophilic cells at interphase in exposed and control groups in 1954-1956

Abnormalities	Exposed group	Comparison group
Binuclear cells	0.3±0.6[b] (0-2)[c]	0.2±0.4 (0-1)
Giant cells	0.5±1.3 (0-7)	0.1±0.3 (0-1)
Fragmentation of nuclei	2.1±3.3 *p= 0.0021* [a] (0-13)	0.3±0.8 (0-3)
Karyolysis	0.5±2.0 (0-10)	0
Total lethal abnormalities	3.3±3.7 *p=0.0001* (0-13)	0.5±0.8 (0-3)
Pyknosis of nuclei	6.4±10.9 *p=0.0002* (0-47)	0.4±1,0 (0-4)
Total abnormalities	9.8±12.8 *p=0.0002* (0-57)	0.9±1.1 (0-4)

[a] Based on Mann-Whitney *U*-test comparing the exposed and control groups; [b] mean ± standard deviation; [c] minimum-maximum.

It was noted that during the maximum exposures the RBM erythrocytic lineages in the exposed group members contained a considerably larger number of abnormal cells and cells with lethal defects (interchromosomal bridges, binuclear cells, micronuclei) than those in the control group (Table 6). Of special importance was high frequency of erythrokaryocytes linked by cytoplasmatic bridges.

Table 6. Frequency of abnormalities per 1,000 erythrokaryocytes in the exposed subjects versus the controls in 1954 – 1956

Abnormalities	Exposed group	Control group
Interchromosomal bridges	0.2±0.4[b] (0-2)[c]	0
Binuclear cells	5.9±3.4 (0-15)	4.7±2.5 (1-11)
Micronuclei	2.6±2.1 *p=0.0003*[a] (0-8)	0.8±0.8 (0-3)
Total lethal abnormalities	8.5±4.3 *p=0.0105* (0-20)	5.5±2.7 (1-12)
Cytoplasmic bridges	20.3±12.0 *p<0.0001* (3-47)	3.8±2.9 (1-10)
Pycnosis of nuclei	0.13±0.4 (0-2)	0
Total abnormalities	30.0±12.9 *p<0.0001* *(0-62)*	9.1±3.4 (4-16)

[a] Based on Student's *t*-test comparing the exposed and control groups except for interchromosomal bridges, pyknosis of nuclei and micronuclei for which Mann-Whitney *U*-test was used; [b] mean ± standard deviation; [c] minimum-maximum.

Thus in 1954-1956, the reaction of the RBM granulocytopoiesis to chronic radiation exposure was manifested by delayed maturation of neutrophilic cells. Differentiation of granulocytes was delayed at the phase of maternal myelocyte, and in parallel to that the number of blasts was increasing. In general, no changes were noted in the proliferative activity and mitotic index of granulocytes. Persons chronically exposed to radiation also demonstrated an increased rate of proliferation and accelerated maturation of erythrokaryocytes. The number of erythroblasts was reduced.

The studies of the exposed individuals made it possible to show that during the period of maximum exposures a significant proportion of dividing RBM neutrophils and erythrokaryocytes manifested morphological signs of abnormal mitotic processes (chromosomal stickiness), a phenomenon that was reflected in an increased frequency of chromosome stickiness ("linked metaphases"), and lethal interphase abnormalities. However, the role played by abnormalities of RBM cell mitosis in the induction of post-radiation hemopoiesis inhibition has not been studied in sufficient detail so far.

Peripheral Blood Cell Composition during the Period of Maximum Radiation Exposure (1951-1956)

It should be noted that during the maximum radiation exposure the changes going on in the bone-marrow were reflected in the peripheral blood cell composition. Blood leukocyte and thrombocyte counts were highly dependent on RBM dose-rates. As is evident from Table 7, the peripheral blood of the chronically exposed group was characterized by decreased counts of leukocytes and their main populations (lymphocytes and neutrophils), and thrombocytes. The lowest values of these parameters were observed during the first 3 years of exposure (1951-1953) when the highest RBM dose-rates were registered. During the same time interval, the exposed group manifested decreased frequency of peripheral blood monocytes and basophils. The count of erythrocytes was within the normal limits in the exposed group for both genders, and it did not show any significant differences from that in the control group. The stability of peripheral blood erythrocyte counts in the exposed group was probably maintained from 1951 to 1956 owing to adequate compensatory processes in the bone marrow.

Decreased counts of leukocytes noted over the initial 3 years of radiation exposure were associated with decreased counts of both lymphocytes and neutrophils, and in the ensuing years only with decreased levels of neutrophils. In 1954-1956 an increase in stab neutrophils was noted in differential leukocyte counts.

In the exposed group a significant ($p<0.0001$) increase in the frequency of moderate leucopenia (leukocyte counts being from 3.3×10^9/L to 4.0×10^9/L) was observed during the period of maximum exposures. It was registered in 11.1% of the exposed group 1-3 years after the start of exposure, and diminished to 7.3% in 1954-1956. A marked decrease in blood leukocytes (< 3.3×10^9/L) registered in 2.6% of the exposed group in 1951-1953 decreased to

1.5% in 1954-1956. One per cent of the control group manifested only moderate leucopenia.

Table 7. Distribution of peripheral blood cell composition (mean±standard deviation) in exposed and control group in the period of maximum radiation exposure (1951-1956)

Parameter	Exposed group		Control group
	1951-1953	1954-1956	
Leukocytes ($\times 10^9$/L)	5.3±1.4 [b,e]	5.7±1.5 [e]	6.4±1.4
Thrombocytes ($\times 10^9$/L)	190.4±42.8 [e]	223.3±63.9 [e]	256.1±65.3
Neutrophils ($\times 10^9$/L)	3.0±1.0 [e]	3.1±1.2 [e]	3.5±1.1
Lymphocytes ($\times 10^9$/L)	1.8±0.6 [e]	2.0±0.7	2.2±0.8
Stab neutrophils (%)	8.0±5.2	8.9±5.8 [e]	7.3±4.4
Segmented neutrophils (%)	48.3±9.0 [b]	43.6±11.4 [d]	46.4±10.5
Erythrocytes ($\times 10^9$/L)			
Male	4.3±0.5	4.5±0.5	4.6±0.4
Female	4.0±0.4 [b]	4.2±0.5	4.2±0.4
Monocytes (%)	6.3±2.8 [d]	7.6±3.7	7.3±3.7
Eosinophils (%)	3.6±2.7	3.2±2.6	3.7±2.6
Basophils (%)	0.18±0.47 [c]	0.37±0.68 [c]	0.28±0.64

[a] *p*-value based on Student's *t*-test comparing the exposed and control groups in the respective periods; [b] $p<0.05$; [c] $p<0.01$; [d] $p<0.001$; [e] $p<0.0001$.

Over the initial 6 years of chronic radiation exposure, a reduction in the average leukocyte counts in the exposed group was attributed to an increased number of individulas with leucopenia, and to the fact that leukocyte counts were approaching the lower limit of the physiological norm. The distribution of leukocyte counts in the exposed group depicted markedly asymmetric curves for the years 1951-1953 and 1954-1956, with a shift towards lower leukocyte counts compared to the control group.

A significant (p=0.0205) increase in the frequency of lymphopenia in the exposed group compared to the control group was only observed during the initial 3 years after the start of chronic exposure, without any increase in the frequency of manifest lymphopenia (lymphocyte $<0.8\times 10^9$/L). Neutropenia was significantly (p=0.0004) more frequently observed in the exposed group than in the control group 4-6 years after the start of exposure: in 1.6% and 13.8% of the exposed group, neutrophil counts were below 1.3×10^9/L (pronounced neutropenia) and were within the range of 1.3-1.9 $\times 10^9$/L (moderate neutropenia), respectively. In the control group, only 7.2% of individuals were noted to have developed moderate neutropenia.

Thrombocytes in the exposed group were significantly reduced in 1951-1956, especially during the first 3 years compared to the control group (Table 7). Those with thrombocyte counts in the range of 150-180 $\times 10^9$/L (moderate thrombocytopenia) and those with the counts below 150×10^9/L (pronounced thrombocytopenia) were significantly more frequent in the exposed group than in the control group ($p<0.0001$ and $p=0.0003$, respectively). Besides, with a decrease in exposure dose-rate to RBM overtime, the proportion of individuals with moderate thrombocytopenia (27.6% in 1951-1953 and 13.7% in 1954-1956 in the exposed group, and 4.7% in the control group), and manifest thrombocytopenia (14.4% in 1951-1953 and 10.1% in 1954-1956 in the exposed group, and 4.1% in the control group) was decreasing.

The distribution of thrombocytes in the exposed group in 1951-1953 and 1954-1956 shifted to the left from that in the control group, and the shift was most pronounced in 1951-1953.

Erythrocyte counts in the exposed group were on the whole normal in both genders during the initial 6 years of exposure, and did not differ from those registered in the control group. Frequency of individuals with reduced erythrocyte counts (erythropenia) and that in individuals with increased counts (erythrocytosis) were both similar in the exposed and the control groups.

Thus, in the early years of chronic exposure (1951-1953) at RBM dose-rates reaching 2.03 Gy/yr, a reduction in leukocyte (neutrophil and lymphocyte) and thrombocyte counts was noted. The peripheral blood cell composition started to recover after dose rate had decreased to 0.52 Gy/yr (1954-1956). Besides, the findings of hematological examinations performed during the period of maximum radiation exposures indicated an increased frequency of peripheral blood leucopenia (neutropenia and lymphopenia) and thrombocytopenia in 1951-1953, with a partial recovery of peripheral blood parameters in 1954-1956 following a reduction in exposure dose rate to 0.52 Gy/yr and lower.

DYNAMICS OF BLOOD CELL COMPOSITION IN PERSONS CHRONICALLY EXPOSED TO RADIATION IN 1951-2008

Throughout the follow-up period, erythrocyte counts in men and women comprising the exposed group were maintained at a normally stable level. Leukocyte, neutrophil, lymphocyte and thrombocyte counts depended on RBM dose-rate. As is indicated in the previous section, the lowest counts of leukocytes, neutrophils, lymphocytes and thrombocytes were observed during the period of maximum radiation exposure, while after the dose rates to RBM had decreased, a marked tendency towards normalization of the counts of these cells became apparent as is demonstrated by the linear trend.

The most significant reduction in leukocyte counts was observed during the first decade after the start of exposure (Figure 5A). The normalization of leukocyte counts proved to be a lengthy process. Leukocyte counts observed over 35 years after the start of exposure were significantly lower in the exposed group than in the control group. Since 1986, no statistically significant difference has been observed in the average leukocyte counts between the exposed and control groups. Over the entire follow-up period, the proportion of individuals with moderate leucopenia (Figure 5B) was higher in the exposed group (4-12%) than in the control group (1.0-1.2%). Individuals with manifest leucopenia were registered in the exposed group over the period from 1986 to 2008, while no such cases were seen in the control group up to 2000, and in 2008 no more than 1.7 % were registered. In the late period after the start of chronic radiation exposure (60 years later) only a few individuals with RBM doses reaching 6.69 Gy (mean dose of 0.89 Gy) had marked peripheral blood leucopenia.

The duration and completeness of normalization of leukocyte count were to a significant degree determined by the extent of hemopoietic function inhibition during the period of maximum radiation exposure. The longest recovery period and increased frequency of leukopenia was observed in persons who had shown marked hemopoiesis inhibition during the initial period of exposure. It should be noted that individual characteristics, such as gender and age at exposure, have exerted a significant modifying effect on the recovery of leukocyte counts in chronically exposed individuals. In women, the time required for normalization of leukocyte counts was longer than that in men. The hemopoietic system in younger persons (up to 20 years) was noted to have higher capacity for recovery of leukocyte counts than in older persons exposed to similar RBM doses. The presence of inflammatory diseases also led to a significantly slower hemopoiesis recovery.

Reduction in neutrophil counts persisted over the first 10 years of exposure, which caused a decrease in leukocyte counts (Figure 6A). The minimum level of neutrophilic granulocytes was noted in 1956-1960. Besides, a decrease was noted in the composition of the leukocyte population due to an increase in stab cells, the phenomenon being most manifest in 1954-1956.

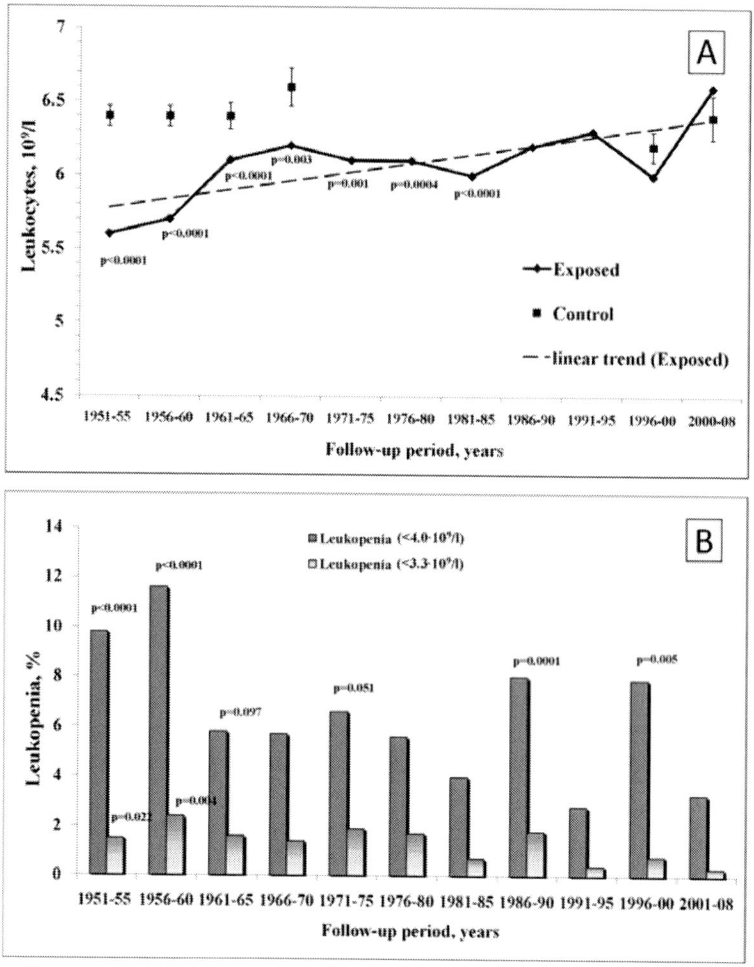

Figure 5. Dynamics of average peripheral blood leukocyte counts in chronically exposed subjects (A); frequency of leucopenia (B) during the period from 1951 through 2008. The dashed line indicates a linear trend for exposed individuals.

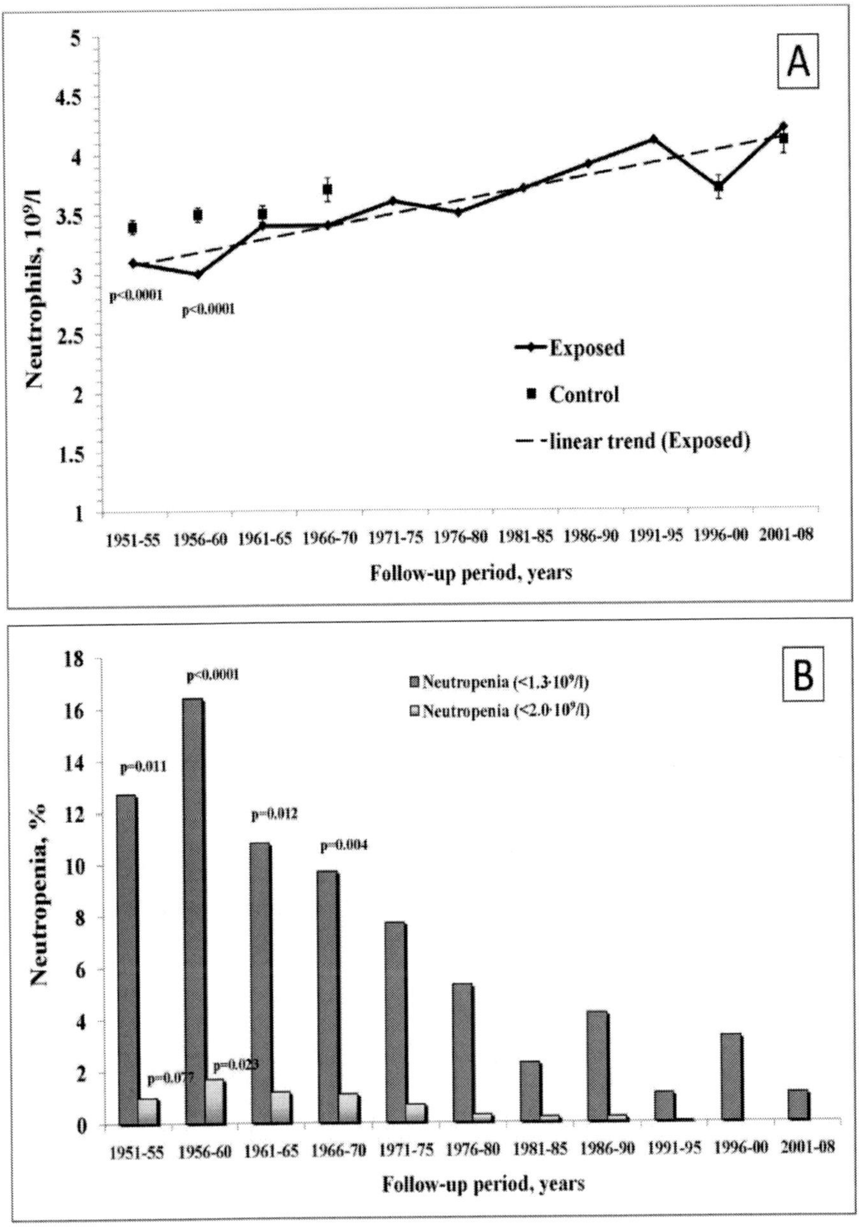

Figure 6. Dynamics of average peripheral blood neutrophil counts in chronically exposed subjects (A); frequency of neutropenia (B) during the period from 1951. The dashed line indicates a linear trend for exposed individuals.

From 1960 onward, the level of neutrophils kept increasing; however, no difference from the values registered in the control group was traced. The proportion of individuals with neutropenia was the highest over the first 10 years of exposure, reaching 16% in 1960, the 10th year of exposure (Figure 6B). From 1976 onward, the incidence of neutropenia (neutrophil $<2.0\times10^9$/L) observed in 1.1-5.3% of the exposed group did not exceed the incidence seen in the control group (6%).

Over the period from 1951 through 2008, 54 cases of manifest stable neutropenia (ICD-9 code 4.288.0) were registered among exposed TRC members without malignant tumors, infectious, parasitic and other diseases which are associated with reduced counts of neutrophilic granulocytes in the blood. In more than half of the cases, neutropenia was observed during the period of maximum radiation exposure (1951-1956). In the remainder proportion of the patients, neutropenia was revealed in the course of medical examinations conducted in later years.

The distribution of patients with neutropenia by dose to RBM is presented in Table 8. It is evident that following an increase in the cumulative RBM dose from 0.1 to 2.62 Gy, odds ratio (OR) of neutropenia development increases relative to that estimated for the controls. After the cumulative dose of 0.5 Gy is reached, it becomes 6.62 times higher (p=0.012) than the respective value obtained for the controls. As a result of further increase in the RBM dose, the OR of development of peripheral blood neutropenia in exposed individuals is 13.79 times higher than that of the development of this disorder in the controls (p<0.0001).

Table 8. Distribution of cases of peripheral blood neutropenia among persons exposed on the Techa River by dose categories

Dose range, Gy	Mean dose to RBM, Gy	Number of cases	OR	95% CI*	p-value
0-0.1 (control)	0.03	2	1	-	-
0.1-0.25	0.18	4	2.76	0.51-15.09	*p=0.418*
0.25-0.5	0.36	6	2.55	0.51-12.65	*p=0.404*
0.5-0.8	0.64	10	6.62	1.45-30.26	*p=0.012*
>0.8 (up to 2.62)	1.39	32	13.79	3.3-57.59	*p<0.0001*

*CI: Confidence Interval.

A linear dependence of OR of neutropenia development on cumulative RBM doses was noted in exposed persons (Figure 7).

The data presented above allow an inference about a significant RBM dose-dependent increase in the probability of peripheral blood neutropenia in persons chronically exposed to radiation due to intakes of Sr^{90} isotopes and external exposure to γ-radiation at RBM doses above 0.5 Gy.

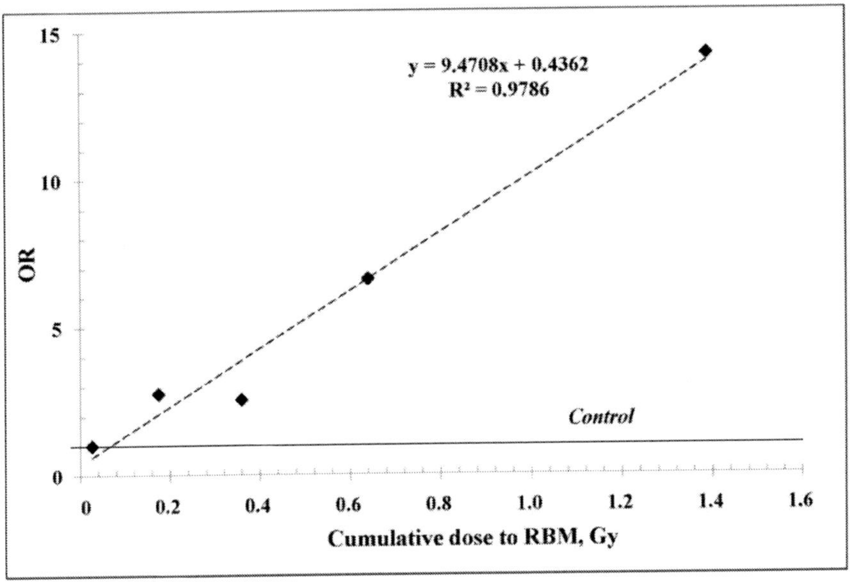

Figure 7. Dependence of the odds ratio of peripheral blood neutropenia development on cumulative dose to RBM in exposed persons versus the controls.

The minimum level of lymphocytes relative to the control group was observed during the initial 5 years of exposure when the highest RBM dose-rates were observed (Figure 8A). In the ensuing years, no clear-cut dynamics was noted in lymphocyte counts. Although the incidence of lymphopenia registered during the overall follow-up period in the exposed group did not, as a rule, exceed that found in the control group (Figure 8B), the mean counts of lymphocytes in the exposed group were periodically found to be lower than those in the control group.

The dynamics of thrombocyte counts depended on dose-rate of BM exposure (Figure 9A). A maximum decrease in the thrombocyte counts as well as in the counts of other blood cells was registered over the period characterized by the highest dose-rates. The level of thrombocytes was restored to normal 10 years after the start of exposure, and it was associated with a significant reduction in the dose-rate of RBM exposure. From 1961 onward, the mean levels of these cells in the exposed group did not differ

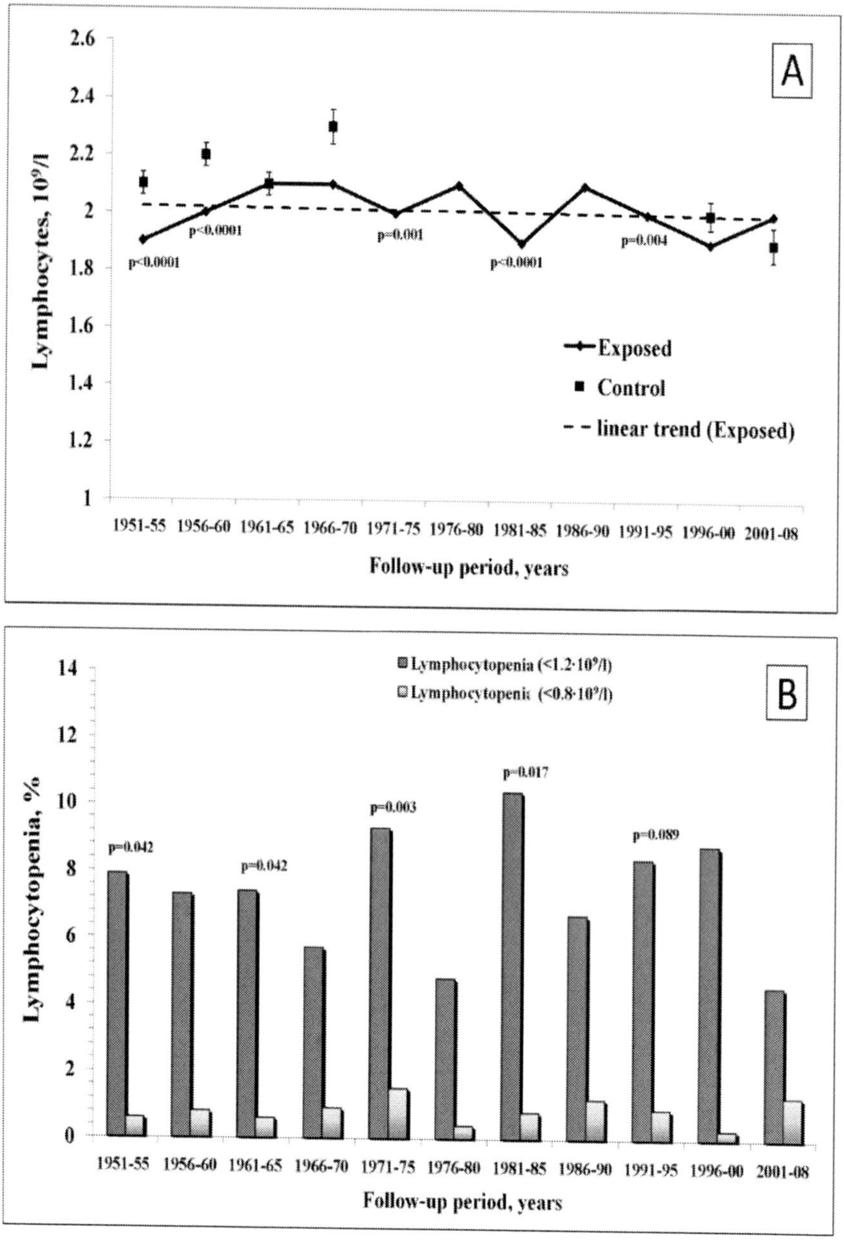

Figure 8. Dynamics of average peripheral blood lymphocyte counts in chronically exposed subjects (A); frequency of lymphopenia (B) during the period from 1951 through 2008. The dashed line indicates a linear trend for exposed individuals.

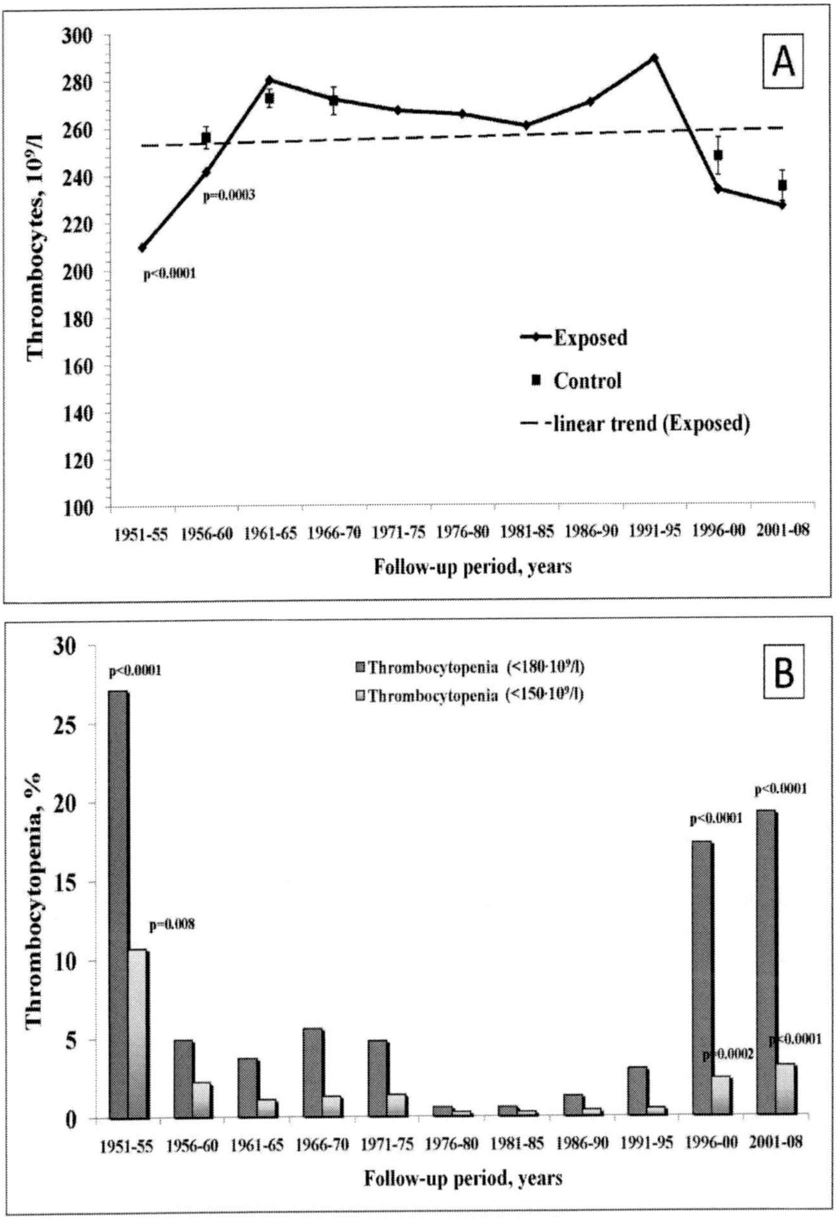

Figure 9. Dynamics of average peripheral blood thrombocyte counts in chronically exposed subjects (A); frequency of thrombocytopenia (B) during the period from 1951 through 2008. The dashed line indicates a linear trend for exposed individuals.

significantly from the respective mean levels in the control group, and the incidence of thrombocypenia registered in the exposed group in 1956 and later did not exceed that in the control group for all time periods (Figure 9B). However, in 1996-2000 the frequency of thrombocytopenia increased significantly again, having reached 17.3-19.2%.

In spite of the fact that the frequency of thrombocytopenia in the controls increased too during that period, the differences between the groups compared were statistically significant. A manifest thrombocytopenia was revealed in 2.4-3.2 % of exposed persons versus 0.9-1.2% in the control group members. The occurrence of excess cases of thrombocytopenia at late time after the onset of chronic exposure is not quite clear, but it may evidently be associated with involutional changes in the megakaryocytic cell series.

Thus, according to the above-indicated data, hemopoiesis inhibition observed in residents of the villages located on the banks of the contaminated Techa River and manifested by a reduced number of peripheral blood cells resulted from a higher radiosensitivity of the two hemopoietic cell lines: granulocytic and megakaryocytic. It should be noted that the radiation-induced reduction in cellularity of the granulocytic series of hemopoiesis proved sufficiently stable over time. The degree of hemopoiesis inhibition is determined not only by the presence of cytopenia, but also by the involvement in the process of several hemopietic cell lines, including relatively radioresistant ones.

Figure 10 shows the frequency of peripheral blood cytopenia resulting from a simultaneous inhibition of the granulocytic (the number of neuthrophilic granulocytes $<2.0 \cdot 10^9$/L) and megakaryocytic series (thrombocytes $<180.0 \cdot 10^9$/L) in comparison with the frequency of leucopenia. It can be seen from the plot that the combination of reduced counts of granulocytes and thrombocytes was registered rather infrequently, and it did not exceed 3.0% of the total cases of cytopenia. Significant differences in this parameter from the controls were only noted during the first decade of radiation exposure on the Techa River.

The incidence of peripheral blood cytopenia associated with a simultaneous reduction in the number of granulocytes, megakaryocytes and erythrocytes in the peripheral blood was observed to be no higher than 1.6% during the follow-up period (Figure 11). Significant differences from the control group were only observed during the first six years of the exposure when the bone marrow dose rates reached 2.03 Gy/yr. Cases of combined peripheral blood granulo-thrombo- and erythropenia among the controls that were only observed in elderly and old members of the group at late time, accounted for 0.5%.

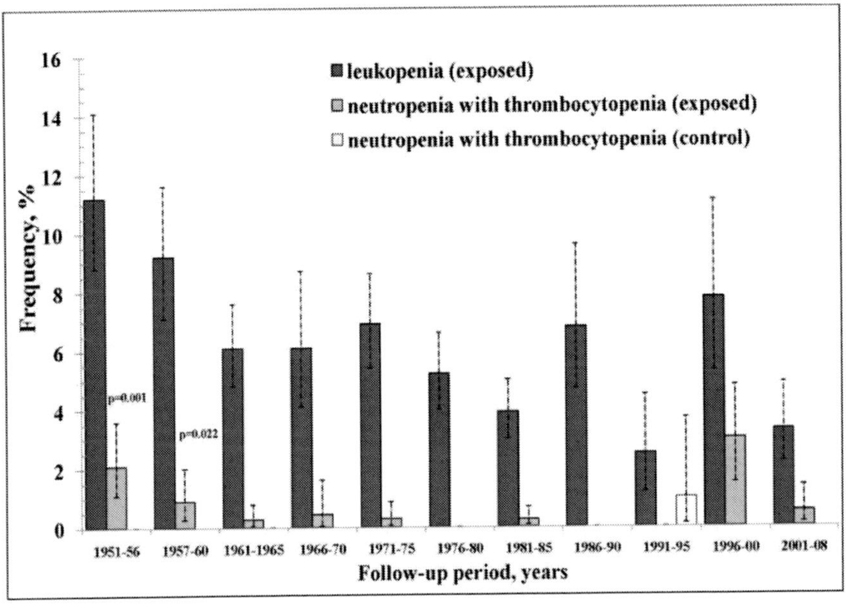

Figure 10. Dynamics of changes in peripheral blood in cases of combined granulocytopenia and thrombocytopenia in exposed subjects *vs* controls (1951-2008). The dashed lines indicate 95% CI.

Thus, most of the cases of cytopenia registered among the exposed individuals were due to reduced counts of leucocytes during the first decade after the onset of exposure. In the exposed persons, inhibition of hematopoiesis which was evaluated on the basis of decreased peripheral blood leucocyte counts was responsible for a decrease in the production of neutrophilic granulocytes in the bone marrow, and to a much lesser degree it could have resulted from a decrease in the content of other types of the white cells, e.g., lymphocytes, etc. Peripheral blood granulocytopenia observed during the period of maximum radiation exposure was determined by a delay in the differentiation of bone-marrow granulocytes at the myelocytic phase, and also by a significant increase in the frequency of lethal damage and defective mitoses. The normal erythrocyte count during the period of maximum chronic exposure was maintained owing to compensatory mechanisms, which involved increased rate of proliferation and maturation of erythrokaryocytes.

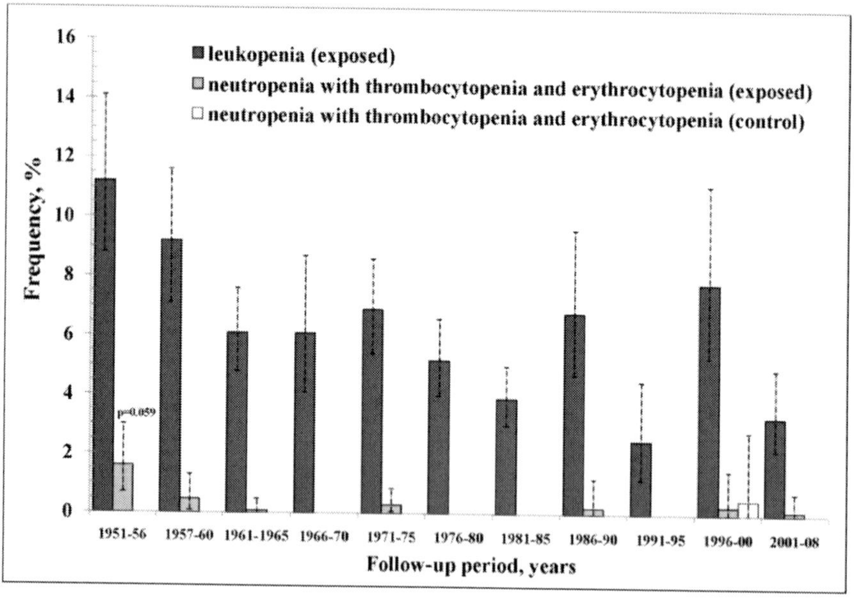

Figure 11. Dynamics of changes in peripheral blood in cases of combined granulocytopenia, thrombocytopenia and erythrocytopenia in exposed subjects and controls (1951-2008). The dashed lines indicate 95% CI.

In earlier years, when radiation exposure was characterized by maximum RBM dose rates reaching 2.03 Gy/yr, cases of cytopenia caused by a combined inhibition of two (granulocytic and megakaryocytic) and three (granulocytic, megakaryocytic and erythrocytic) bone marrow cell lines were identified as well. The maximum incidence of such cases was registered during the third and forth years after the onset of chronic exposure when the maximal dose rates decreased from 2.03 to 0.52 Gy/yr.

LATE EFFECTS OF LDR EXPOSURE

During the 53-year follow-up period (1953-2005) 93 leukemia cases were identified, including 75 for which an incidence data could be determined and 18 which were identified solely from death certificates. Table 9 provides summary information on the distribution of cohort members, person years, and leukemia cases by gender, ethnicity, and age-at-exposure. The crude rates in this table, which do not make allowance for possible dose effects, suggest that rates increase with age and are slightly higher for men than for women. Cohort members identified as Tartar or Bashkir ethnicity appear to have higher rates than those identified as Slavs.

Table 9. Selected characteristics of leukemia cases and TRC cohort members[a]:1953-2005

Category	People	Person Years	Leukemia Cases		Rate (per 10,000 PY)	
			Any	Non-CLL[b]	Any	Non-CLL[a]
Gender						
Male	12,565	336,660	42	30	1.25	0.89
Female	17,191	495,821	51	40	1.03	0.81
Ethnicity						
Tartar and Bashkirs	5,950	200,594	31	26	1.55	1.30
Slavs	23,806	631,887	62	44	0.98	0.70
Age at exposure						
<20	11,769	397,045	40	34	1.01	0.86
20-40	9,640	296,550	31	20	1.05	0.67
>=40	8,347	138,886	22	10	1.58	1.15
Total	29,756	832,482	93	70	1.12	0.84

[a]Catchment area residents only; [b]leukemias other than chronic lymphocytic leukemia (CLL).

Table 10. Crude leukemia rates by years after first exposure

Years since first exposure (average age)	Person Years	Leukemia Cases		Rate (per 10,000 PY)	
		Any	Non-CLL	Any	Non-CLL
0 – 5 (32)	68,684	5	5	0.73	0.73
- 10 (35)	119,439	6	5	0.50	0.42
- 20 (41)	206,818	23	18	1.11	0.87
- 30 (48)	171,488	18	15	1.05	0.87
- 40 (55)	133,216	12	5	0.90	0.38
40+ (63)	132,836	29	22	2.18	1.66
Total	832,482	93	70	1.12	0.84

Table 10 presents data on the person years and leukemia cases by time since exposure. The crude rates in this table, which do not make allowance for possible dose effects, suggest that rates increase with time after exposure.

More than half of the cases (48 cases) were classified as chronic leukemias, 45% (42 cases) as acute or subacute leukemias, while the type of the leukemia could not be determined for 3% (3 cases) of the cases (Table 11).

Table 11. Distribution of leukemias by histologic type: Techa River Cohort 1953-2005

Type	Incident cases	DCO	Total	Mean age at diagnosis
Acute leukemias				
Myeloid	8	0	8	48
Other specified types [a]	3	0	3	63
Unspecified cell types	19	12	31	52
Chronic leukemias				
Lymphocytic	22	1	23	62
Myeloid	22	3	25	57
Other or unspecified leukemias				
Myeloid	1	2	3	56
Total	75	18	93	55

[a]Includes: one case of acute monocytic leukemia and two cases of acute erythremic myelosis.

Mean age at diagnosis was 48-63, the youngest – for acute myeloid leukemia, the oldest – for chronic lymphatic leukemia. It should be noted that no leukemia cases were reported among cohort members under the age of 10. The first leukemia cases were registered in TRC 13 years later (myeloid and unspecified cell types).

There was a statistically significant linear dependence on dose for non-CLL leukemias ($p < 0.001$). The excess relative risk (ERR) value per Gy was 4.9 with a 95% confidence interval of 1.6 to 14 [11]. There was no indication of non-linearity in the non-CLL leukemia dose response ($p > 0.5$) and the estimated curvature was close to 0 (-0.12 with a 95% CI < -0.5; 17) (Figure 12).

Tests for gender or ethnic differences in the non-CLL dose response provided no indication of effects ($p > 0.5$). Allowing the dose response to depend on attained age, age at exposure, or time since first exposure did not improve the fit of the model ($p > 0.5$) and the estimates of the changes were small.

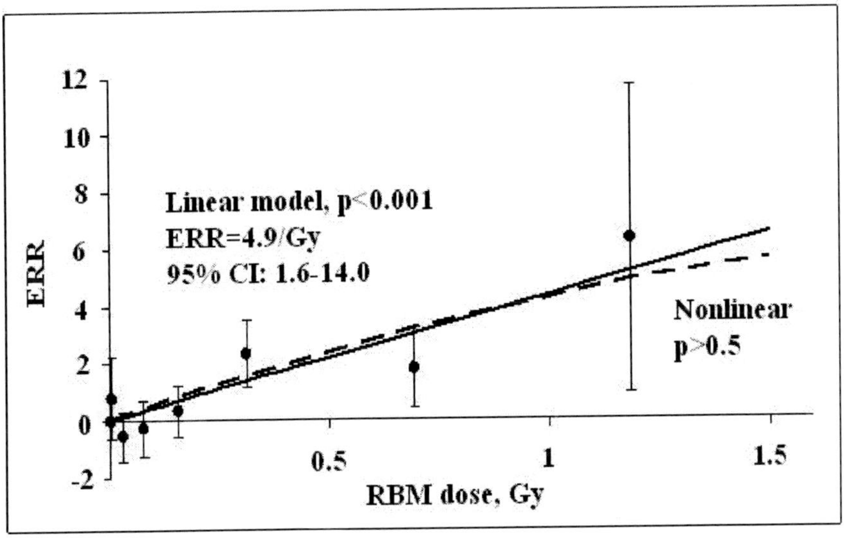

Figure 12. Non-CLL leukemia dose response function: linear model-solid line, linear-quadratic model-dot-dash line, nonparametric model-points with 95% confidence interval.

For total leukemias there was also a statistically significant linear dependence on dose ($p < 0.001$). The ERR per 1 Gy estimate was 2.2 with a 95% confidence interval of 0.7 to 5.4. Using linear-quadratic model did not improve the significance of ERR value ($p > 0.5$).

In contrast, we found no indication of a dose response for CLL ($p > 0.5$) and the ERR per 1 Gy estimate was negative (< -0.2 per 1 Gy) with an upper 95% confidence bound of 1.4.

Table 12 presents the number of person-years and cases in categories of cumulative 2-year–lagged red bone marrow dose. The table also includes estimates of the number of radiation-associated non-CLL and total leukemia cases. It was estimated that 59% (95% CI 32%; 80%) of the non-CLL leukemias and 40% for total leukemias were associated with the radiation-exposure. The proportion of radiation-related cases was higher for those exposed to more than 0.5 Gy.

Over the period of follow-up from 1951 through 2008, 7 cases of aplastic and hypoplastic state of the bone marrow was registered in the TRC of whom 5 were diagnosed with aplastic anemia (ICD-9 code 4.284.8 and 4.284.9) and

2 had severe form of neutropenia with agranulocytosis-type course (ICD-9 code 4.288.0). Reviews of medical records on patients with agranulocytosis did not disclose a clear-cut relationship with autoagressive medication or a past history of systemic autoimmune diseases (systemic lupus erythematosus, rheumatoid arthritis, etc.).

Table 12. Distribution of observed and expected leukemia cases by red bone marrow dose category

Dose category, Gy	PYR	Total leukemia		Non-CLL		CLL Cases [b]
		Cases	Excess [a]	Cases	Excess [a]	
<0.01	103,499	11	0	6	0	5
- 0.1	137,364	6	1	3	1	3
-0.2	171,109	13	4	8	4	5
-0.5	254,262	36	12	31	14	5
-1	140,848	20	15	15	17	5
>=1	25,399	7	5	7	5	0
Total	832,482	93	37	70	41	23
Attributive risk			40%		59%	

[a]excess case estimate from a linear dose response model with no effect modification; [b]excess case estimates not shown since there was no evidence of a dose response (p > 0.5).

Bone marrow hypoplasia and aplasia were registered during the period from 1963 through 2001, with the latency period amounting to 13 years. As can be seen from Table 13, no cases of RBM aplasia and hypoplasia were observed among individuals within the bone marrow dose category of up to 0.5 Gy. After the cumulative dose of 0.5 Gy is reached, the probability of developing aplastic or hypoplastic state of the bone marrow in relation to the controls increases (OR=1.22), but the differences are statistically insignificant (p=0.665). At RBM exposure doses of over 0.8 Gy, a slight tendency to hypoplasia development was observed (OR=1.58; p=0.857).

Thus, the results of epidemiological analyses have provided a clear evidence of an increased non-CLL leukemia risk attributable to chronic radiation exposure. ERR was 4.9 per 1 Gy (95% CI 1.6; 14). Other study for Ukrainian clean-up workers [16] showed the non-CLL ERRs per 1 Gy 2.73 (95% CI < 0; 13.5) and for clean-up workers from Belarus, Russia and the

Baltic countries showed non-CLL ERRs per 1 Gy 5.0 (90% CI -0.04; 57) [17]. TRC non-CLL and total leukemias risk estimates are also consistent with an ERR per 1 Gy for non-CLL mortality in the atomic bomb survivors who received an acute dose of radiation [18].

Table 13. Distribution of cases of RBM hypoplasia among persons exposed to radiation on the Techa River by dose categories

Dose range, Gy	Mean dose to RBM, Gy	Number of cases	OR	95% CI	p-value
0-0.01 (Control)	0.004	1	1	-	-
0.01-0.5	0.25	0	0	-	-
0.5-0.8	0.64	2	1.22	0.11-13.46	p=0.665
>0.8 (up to 2.23)	1.36	4	1.58	0.23-16.92	p=0.857

CONCLUSION

Long-term exposure of the RBM to LDR ionizing radiation, including external exposure to γ-radiation and intakes of ^{90}Sr from drinking water and consumption of local food products, considerably affected hemopoiesis in residents of the Techa riverside villages. The most pronounced changes in the hemopoietic system were observed in 1951-1953 when the dose-rate of RBM exposure reached 2.03 Gy/yr (mean value: 0.24 Gy/yr). At that time, a significant number of the exposed group members were noted to manifest decreased counts of neutrophilic granulocytes, lymphocytes and thrombocytes. As a result of a reduction in the exposure dose-rate (mean value: 0.07 Gy/yr; individual values not exceeding 0.52 Gy/yr) in 1954-1956, leukocyte and thrombocyte counts showed a tendency towards normalization.

The changes observed in the peripheral blood cell composition reflected the reactions of different RBM series to chronic radiation exposure. The stability of normal erythrocyte counts registered in residents of the Techa riverside settlements was maintained owing to adequate compensatory changes in the erythroid lineage of the bone marrow: intensified proliferation rate and increased rate of erythrokaryocyte differentiation. Hyperplasia of the erythroid cell lineage observed during the period of maximum radiation exposure indicates an expanded potential of erythpoiesis. Comparison of the number of cells between those undergoing mitosis and those with lethal damage showed that the proportion of erythrokaryocytes with lethal aberrations in the erythroid lineage of the exposed group (8.5‰) was more than two-fold lower than that of dividing erythrokaryocytes (18.3‰). The situation in the control group was similar: the number of mitoses was essentially higher than that in cells with aberrations. The above-mentioned data lead to an assumption that in chronic exposure (RBM dose-rate ≤0.52 Gy/yr) radiation-induced lethal damage to erythrokaryocytes is compensated for by intensified proliferative activity in the pool of proliferating-maturing cells of the erythroid lineage and by their accelerated maturation, ensuring the numerical stability of erythrocytes in the peripheral blood.

On the other hand, the state of the pool of proliferating-maturing RBM neutrophils during the period of maximal exposure indicated that the alteration process prevailed over repopulation which was reflected in the peripheral blood cell composition. Reduction in the leukocyte counts to 2.2×10^9/L, and in neutrophil counts to 0.9×10^9/L observed in some of the chronically exposed individuals can be attributed to inefficiency of the compensatory-recovery processes in the bone marrow granulocytopoiesis. Thus, the mean number of cells with lethal defects identified in the neutrophilic lineage (3.3‰) was comparable to that of cells undergoing mitotic processes (3.4‰), while the level of bone marrow neutrophils in the state of mitosis observed in the control group (3.7‰) was much higher than that of cells with lethal aberrations (0.5‰).

Another factor bringing about a deficit of segmented neutrophils in the peripheral blood is a delay in maturation of neutrophilic granulocytes in the bone marrow (primarily at the myelocytic phase). It was revealed based on the findings of the studies of myelopoiesis performed for a number of exposed individuals that the content of mature neutrophils persisted was reduced both in bone marrow and in blood in spite of the increased proliferation of neutrophilic granulocytes in bone marrow. At the same time, the neutrophil maturation index was found to be increased. It can be concluded, consequently, that the mechanisms responsible for compensation of the cell loss were not effective enough to lead to adequate repopulation of RBM neutrophils, which prevented a sufficient number of these cells from entering the blood flow. Thus, for a long time period, granulocytopoiesis failed in adequately compensating for cell loss resulting from radiation exposure, and physiological regeneration in exposed individuals.

Normalization of peripheral blood count occurred within different time periods after the start of exposure. Characteristically, lymphocyte, thrombocyte and neutrophil counts were restored to normal 4-6, 7-10 and 12-17 years later, respectively. The longest time was required for the entire population of circulating leukocytes to recover: 8% of the exposed individuals manifested reduced functional activity of the leukocytic lineage late after the start of exposure (50 years later). Thus, the data from this study are indicative of a long duration of the leukocyte count recovery process under chronic radiation exposure of RBM at doses reaching 6.69 Gy.

The factors that modify the process of blood cell composition recovery after chronic exposure include gender, age at exposure and the degree of hemopoiesis inhibition during the period of maximum exposure. At comparable RBM doses normalization of leukocyte counts in women was observed to occur later compared to that in men. In individuals exposed to similar RBM doses, recovery of leukocyte count occurred earlier in those aged 20 years or younger, than in older persons. Recovery period was noted to be prolonged in individuals with a more pronounced initial hemopoiesis inhibition (leukocyte count $< 3.3\times10^9$/L).

In most cases, spontaneous (without treatment) hemopoiesis recovery was observed. However, excess cases of leukemia were registered among residents of the Techa riverside villages. The latency period for development of leukemia was at least 10 years. Our current results demonstrate a statistically significant linear dependence ($p<0.001$) of the excess relative risk of total leukemias and of non-CLL leukemia incidence on cumulative dose to red bone marrow with almost 60% of the non-CLL cases and 40% of total leukemias in the cohort attributable to radiation exposure. The ERR per Gy of 4.9 for non-CLL and 2.2 per 1 Gy for total leukemias in this analysis was lower, albeit still statistically compatible with the estimate from an early mortality analysis [19]. The difference in the ERR estimates is primarily due to the extended follow-up period (6 additional years) and the addition of many incident cases (about 50% more cases) comparing with mortality analyses. The suggestion that the reduction in risk is related to the additional cases is supported by the similar risk estimate (ERR per Gy 4.6; 95% CI: 1.7; 12.3) found in a case-control study of 83 incident leukemia cases conducted in the TRC before TRDS-2000 was available [20].

Thus, chronic LDR exposure at dose rates reaching 2.0 Gy/yr may bring about deficiency of compensatory-adaptive mechanisms for inadequacy of the adaptive capacity of the hemopoietic system at early time of exposure [3, 4, 21]. It should be noted that the impairment of the hemopoietic function observed at late time after exposure can be associated not only with the development of pathological hyperplastic processes (leukemias), but also with organic changes in the bone marrow (e.g., RBM hypoplasia and vascular changes) which hamper the physiological regeneration of hemopoietic system.

References

[1] Guskova, A.K., and Baysogolov, G.D. (1971). *Radiation Disease in Human.* Moscow, RU: Meditsina.

[2] Konoplyannikov, A.G. Radiobiology of Stem Cells. Moscow, RU: Energoatomizdat, 1984.

[3] Seed, T.M, Fritz, T.E., Tolle, D.V., and Jackson, W.E. (2002). Heamopoietic responses under protracted exposures to low daily dose gamma irradiation. *Adv. Space Res, vol. 30*, pp 945-955.

[4] Seed, T.M., Tolle, D.V., and Fritz, T.E. (2002). Haematological responses to chronic irradiation: the past Argonne experience and future AFRRI initiatives. *Br J Radiol Suppl, vol.26*, pp 94-103.

[5] Degteva, M.O., Shagina, N.B., Vorobiova, M.I., Anspaugh, L.R., and Napier, B.A. (2011). Re-evaluation of waterborne releases of radioactive materials from the «Mayak» production association into the Techa river in 1949-1951. *Health Physics*, (in print).

[6] Akleyev, A.V., and Kisselyov, M.F. (Eds). (2002). *Medical-biological and Ecological Impacts of Radioactive Contamination of the Techa River.* Moscow, RU: Medbioextrem.

[7] Akleyev, A.V., and Lyubchansky, E.R. (1994). Environmental and medical effects of nuclear weapon production in the Southern Urals. *Sci. Total Environ, vol. 42,* pp 1-8.

[8] Smirnova, L.G., and Kost, E.A. (Eds). (1960). *Manual on Clinical Laboratory Investigations.* Moscow , RU: Medgiz.

[9] Menshikov, V.V. (Ed.). *Laboratory Methods of Examination in Clinical Setting.* Moscow , RU: Medicine, 1987.

[10] Sokolov, V.V., and Gribova, I.A.. (1972). *Hematological Characteristics of Healthy Human*. Moscow , RU: Medicine .

[11] Krestinina, L.Y., Preston, D.L., Davis, F.G., Epifanova, S.B., Ostroumova, E.V., Ron, E., and Akleyev, A.V. (2010). Leukemia incidence among people exposed to chronic radiation from the contaminated Techa River, 1953-2005. *Radiat. Environ. Biophys, vol. 49,* pp 195–201.

[12] Kossenko, M.M., Thomas, T.L., Akleyev, A.V., Krestinina, L.Yu., Startsev, N.V., Zhidkova, C.M., Vyushkova, O.V., Hoffman, D.A., Preston, D.L., Davis, F., and Ron, E. (2005). The Techa River Cohort: Study Design and Follow-up Methods. *Radiat. Res, vol. 164,* pp 591–601.

[13] Preston, D.L., Lubin, J.H., Pierce, D.A., and McConney, M.E (1993). Epicure Users Guide. Washington, US: Hirosoft International Corporation, Seattle.
[14] Degteva, M.O., Vorobiova, M.I., Kozheurov, V.P., Tolstykh, E.I., Anspaugh, L.R., and Napier, B.A. (2000). Dose reconstruction system for the exposed population living along the Techa River. *Health Phys, vol. 78,* pp 542–554.
[15] Clayton, D., and Hills, M. (1993). Statistical Methods in Epidemiology. New York, US: Oxford University Press.
[16] Romanenko, A.Y., Finch, S.C., Hatch, M., Lubin, J.N., Bebeshko, V.G., Bazyka, D.A., Gudzenko, N., Dyagil, I.S., Reiss, R., Bouville, A., Chumak, V.V., Trotsiuk, N.K., Babkina, N.G., Belayev, Y., Masnyk, I., Ron, E., Howe, G.R., and Zabloska, I.B. (2008). The Ukrainian-American study of leukemia and related disorders among Chernobyl cleanup workers from Ukraine: III. Radiation risks. *Radiat. Res, vol. 170,* pp 711-720.
[17] Kesminiene, A., Evrard, A.S., Ivanov, V.K., Malakhova, I.V., Kurtinaitis, J., Stengrevics, et al (2008). Risk of hematological malignancies among Chernobyl liquidators. *Radiat. Res,* vol. *170,* pp 721-735.
[18] Preston, D.L., Pierce, D.A., Shimizu, Y., Cullings, H.M., Fujita, S., Funamoto, S. and Kodama, K. (2004). Effect of recent changes in atomic bomb survivor dosimetry on cancer mortality risk estimates. *Radiat. Res, vol. 162,* pp 377-389.
[19] Krestinina, L.Y., Preston, D.L., Ostroumova, E.V., Degteva, M.O., Ron, E., Vyushkova, O.V., Startsev, N.V., Kossenko, M.M., and Akleyev, A.V. (2005) Protracted radiation exposure and cancer mortality in the Techa River Cohort. *Radiat. Res, vol. 164*, pp 602– 611.
[20] Ostroumova, E.V., Gagniere, B., Laurier, D., Gudkova, N.V., Krestinina, L.Y., Verger, P., Hubert, P., Bard, D., Akleyev, A.V., Tirmarche, M., and Kossenko, M.M. (2006). Risk analysis of leukaemia incidence among people living along the Techa River: a nested case-control study. *J. Radiol. Prot, vol. 26.* pp 17–32.
[21] Pesternikova, V.S., and Okladnikova, N.D. (2004). Assessment of bone marrow hemopoiesis in patients with chronic radiation sickness after 40 years of follow-up. *Radiation Safety Issues, vol. 4,* pp 41-45.

In: Radiation Exposure
Editors: D. Balenovic and E. Stimac
ISBN 978-1-62100-869-9

Chapter 4

The Current State of Terrestrial Ecosystems in the Eastern Ural Radioactive Trace

V. N. Pozolotina,* I. V. Molchanova, L. N. Mikhaylovskaya, E. V. Antonova, and E. N. Karavaeva

Institute of Plant and Animal Ecology, Ural Division of Russian Academy of Sciences, Yekaterinburg, Russian Federation

Abstract

A large-scale nuclear accident took place at the Production Association "Mayak" in the Southern Urals in 1957. About 74×10^{15} Bq of radioactive substances were released into the atmosphere, which resulted in the contamination of a vast area, the Eastern-Ural Radioactive Trace (EURT). In this case, the main long-lived radionuclide was ^{90}Sr. In 1967, this area was subjected to a secondary contamination resulting from radioactive sediments transported by wind from the Mayak technological reservoir, Karachay Lake. The ^{137}Cs was predominant in the composition of this contamination. Two contamination range areas were established: impact and buffer zones. Within EURT area the radionuclide contamination density decreases with the distance from the accident location and is fairly described as a power function. The specific activity

*E-mail address: pozolotina@ipae.uran.ru.

of radionuclides in the herbaceous plants depend on the density of soil contamination. However, transfers coefficients for plants of impact zone were, as a rule, lower than that for background one. Calculations indicated that inpresent stock of the radionuclides in EURT soils is 570×10^{12} Bq for ^{90}Sr, 66×10^{12} Bq for ^{137}Cs and about 2×10^{12} Bq for ^{239,240}Pu. The stock in the impact zone consist of 75% of ^{90}Sr, 70% of ^{239,240}Pu and only 23% of ^{137}Cs from their total amount within EURT zone. Most of ^{137}Cs stock is located in the buffer area where it was received as a result of the Karachay accident. Doze loads upon some herbaceous plants were evaluated. In the impact zone, they exceed background level on 1–3 orders of magnitudes. The aspects of flora diversity within the EURT were studied. The natural growth in the EURT zone is a complex of synanthropic and semi-natural cenosises on the different stages of degradation and restoration. Species diversity in the impact zone is very high; some communities have up to 65 species, it is associated with a low anthropogenic load. The current state of phytocenosises in the demolished local villages is still largely determined by the degree of man's impact in the pre-accident period and recultivation activities. The natural ecosystems successfully recovered at a faster pace.

INTRODUCTION

Modern society faces the need to build its activity within the parameters that determine the stability of the biosphere. Under conditions of an increasingly growing technological burden, one should take into account the formation of a new, globally ecological factor – introduction into the biosphere of artificial radionuclides and the ionizing radiation generated by them. Currently, one of the main ways of radionuclides getting into the environment is through the operation of atomic power plants, which is often accompanied by emergency situations.

Such a complicated radio-ecological situation has developed in the Urals region. Here, during the development of the atomic project in the middle of the last century a powerful production association (PA) "Mayak" was established. At this plant on September 29, 1957 a large radiation accident occurred (so-called Kyshtym accident), during which 74×10^{15} Bq of radioactive substances were discharged into the atmosphere. It resulted in the contamination of large area (23000 km^2), named later as the East-Ural radioactive trace (EURT). ^{90}Sr prevailed among long-lived radionuclides in the emergency emissions. Later, in 1967 this same area was exposed to a secondary contamination caused by wind transfer of radioactive bottom sediments from the shores of the Karachay Lake – the technological reservoir of PA “Mayak”. The main contaminant in this case was ^{137}Cs. We have carried out a long-term radio-ecological investigation of the EURT territory in two complementary directions. The first direction includes an assessment of the intensity of radionuclides migration in soils and food chains leading to humans. The second one covers the study of the biological effects on plants under conditions of chronic exposure.

The objectives of this work: 1) ranking of EURT territory in terms of contemporary radionuclide contamination, 2) study of the features of migration of long-lived radionuclides within distinguished zones, 3) evaluation of dose rates, 4) estimation of the state of natural plant communities in the EURT area.

Material and Methods

The territories studied are in the East Ural radioactive trace located within the Trans-Ural forest-steppe, which is characterized by the alternation of steppe meadows, birch, birch and aspen small groves and pine forests [1]. The soil cover is dominated by various subtypes of gray forest soils and chernozem soils of varying thicknesses and degrees of leaching; varieties of brown forest and peaty soils are here as well. Throughout the EURT there are areas with soil cover disturbed by post-emergency rehabilitation measures.

In the coursc of studies at different distances from the epicenter of the accident, plots were selected covering the main types of ecosystems. Soil profiles were laid on each of them for sampling. As a rule, layers of soil samples were taken with 5–10 cm capacity to a depth of 50 cm. The total content of radionuclides in the 50-cm soil layer, rated to the area, was called the contamination density of radionuclides or the stock. In the vicinity of the profiles, samples of above-ground mass of the dominant plant species, forest and meadow grasses were taken. Background areas, where soil and plant samples were also taken, had been located outside the emergency emission zone in 1957, about 30–40 km from the western border of EURT. All samples were dried to air-dry state, and then incinerated at t=450° C. The content of ^{137}Cs in selected samples was determined by a gamma-ray analyzer with a germanium semiconductor detector "Canberra Packard" (USA) with a detection limit of 0.1 Bq. The content of ^{90}Sr was determined by radiochemical method [2]. The procedures were based on the radinuclide leaching by 6 N HCl-solution from preliminary prepared samples. Precipitating of ^{90}Sr in the oxalate form, separating of ^{90}Sr kept in balance with the daughter product of decay – ^{90}Y. Measurement of the prepared specimen of ^{90}Y was carried out with the use of an alpha-beta radiometer (Russia), detection limit of 0.2 Bq. The plutonium content in samples was also determined by the radiochemical method [3]. This method includes leaching isotopes of plutonium by mixing concentrated acids (HNO_3+HCl) from preliminary prepared samples, precipitating ones on ion-exchange resin, then purifying and stripping them from the ion-exchange column in succession by H_2O and 0.5 N HNO_3 – solution, electrodepositing them on disks made of stainless steel. Measurement of prepared specimens was carried out with the use of an Alpha Analyst type of "Canberra Packard" (USA) spectrometer with semiconductor detectors (PIPS) and GENIE-2000 software. The lower limit of detection was 0.001 Bq. The procedural error of the methods did not exceed 20%. For correct assessment of radionuclides in samples collected in different years, the correction for the radioactive decay was applied.

Geobotanical studies were based upon a scale of abundance of species proposed by O. Drude [4]. It identifies the following gradation: un. (unicum) – species is represented in the community by a single specimen, whose projective cover (PC) is 0.1–1% of the area of test area plot; sol. (solitariae) – species is met sporadically or in very small quantities, with PC 1–5%; sp. (sparsae) – species appears in small amounts or occasional dissemination, with PC 6–10%; $cop._1$ (copiosae) – species grows quite profusely, with PC 11–25%; $cop._2$ – species grows abundantly or scattered, with PC 26–50%; $cop._3$ – species grows very abundantly or scattered, but throughout the community, with PC 51–70%; soc. (sociales) – species is a background in this community, its PC can reach 100%. In addition, classification of K. Raunkier [4] was applied, who identified five types of herbaceous plants according to the location of the renewal buds relative to the soil surface: 1) phanerophytes – buds are located high above the ground; 2) chamephytes – buds are close to the surface, but in the snow layer; 3) hemicryptophytes – renewal buds are on the surface level; 4) cryptophytes – renewal buds are in the soil, and 5) therophytes – annuals with wintering seeds.

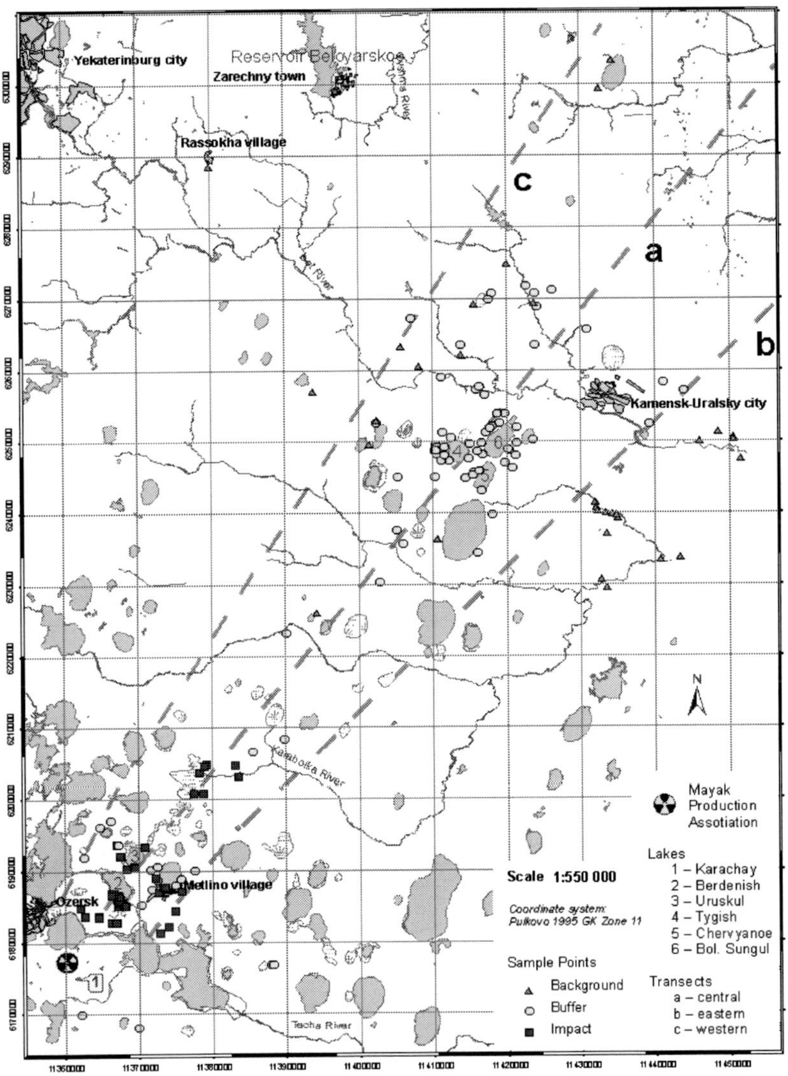

.Figure 1. The sketch-map of the investigation region.

Figure 2. The albino-type chlorophyll mutation in *Cirsium setosum* population.

1.1. The Spatial Distribution of Long-Lived Radionuclides in Soil-Vegetation Covers at EURT

The experience gained in the radio-ecological survey of contaminated areas indicates the need for their ranking according to the content of radionuclides. It facilitates the interpretation of the received data and helps provide a strategy for further study of features of migration and redistribution of radionuclides in soil and vegetation cover. Ranking results are also used to arrange non-uniformly scaled monitoring, determine body burden on living organisms, and justify social protection and rehabilitation measures. The ranking is based on the principle of separation of impact and buffer and background zones. The impact zone is located in the area of the maximum content of radionuclides; the buffer zone is adjacent to the impact zone and includes territory with a gradient of radionuclide contamination falling to background level [5, 6].

EURT territory was ranked according to the above outlined principles. Considering the significant trace length, the gradient of contamination, and the wide variety of soils and ecosystems, we have established three transects on the surveyed territory. Figure 1 shows the central transect (a) coincided with the axis of EURT, eastern (b) and western (c) ones covered the periphery parts of the trace. We agreed that the impact zone was part of EURT area, located near the epicenter of the accident, where the contamination level of ^{90}Sr exceeded 1×10^6 Bq/m^2. The buffer zone was the territory with the gradient of pollution from 1×10^6 Bq/m^2 to the background values, which in selected background plots were 1.6±0.8 kBq/m^2, 5.0±2.0 kBq/m^2 and 0.1±0.05 kBq/m^2 for ^{90}Sr, ^{137}Cs and $^{239,240}Pu$, respectively.

The spatial distribution of radionuclides in the EURT area is shown on Figure 2. It is clear that within the central transect the changes of contamination density of studied radionuclides with increasing distance from the epicenter of the accident are satisfactorily approximated by a power function (R^2 varied within 0.58 – 0.65). In this case, the radionuclides located according to the level of contamination in a row $^{90}Sr > {}^{137}Cs > {}^{239,240}Pu$. On the eastern periphery of the trace the distribution of ^{90}Sr, ^{137}Cs and $^{239,240}Pu$ is subject to the same dependence. However, here the value of ^{90}Sr and ^{137}Cs stocks are almost identical, and with increasing distance come down from 100 kBq/m^2 to 10 kBq/m^2. On the western periphery of the trace the content of ^{137}Cs in soil, regardless of the distance, is kept at a level (5–10) kBq/m^2. Therefore the soil-vegetation cover of areas adjacent to the eastern border of the impact zone is contaminated with ^{137}Cs to a greater extent than that of the western zone. Additional quantities of this nuclide could enter the soil cover of the eastern section from the other sources of pollution. The stocks of $^{239,240}Pu$ are much lower and change from 5 kBq/m^2 to 0.2 kBq/m^2 as the distance from the pollution source grows. We should note the high variability in stocks of radionuclides in the soils of the surveyed sites. This variation can be contributed by the unevenness of the primary precipitation, landscape-geochemical characteristics of radionuclides migration, their additional inflow with regular emissions of PA "Mayak", as well as activities for the rehabilitation of contaminated sites, and the type of ecosystems.

To assess the landscape-geochemical features of migration of radionuclides the geochemical interfaces, characterized by the greatest diversity of types of plant communities, were studied. As an example, the results of one of these studies are listed below. A flow adjacent site was chosen in the impact zone near the Berdenish Lake, 13 km from the epicenter Kyshtym accident. Its territory has a variety of plant communities with native and disturbed during the reclamation operations soil cover. As evident from these data (Table 1), the lowest content of ^{90}Sr, ^{137}Cs and ^{239,240}Pu was found in the frequently flooded banks of the lake. In transit to the watershed area their stock in the soils of selected ecosystems increases. The watershed area includes the reclamated territory of the former village, an undisturbed meadow and small woodland. The area under the forest with its man-made disturbance lead to a decrease in stocks of ^{90}Sr and to some extent stocks of ^{239,240}Pu in the soil.

Such decrease may be determined, on the one hand, by prolonged deposition of pollutants in the perennial organs of woody plants, and on the other hand, by the removal of the upper and most contaminated soil layer during land remediation [7]. Analysis of values of ratio ^{90}Sr/^{137}Cs and ^{90}Sr/^{239,240}Pu in soils show that territories of the bank zone and watershed area are different according to the intensity of radionuclide migration. In the subaqueous ecosystem the lower values of radionuclide ratio reflect the relatively high migration ability of ^{90}Sr, which leads to its self-decontamination [8]. Besides that, decontamination of the bank zone of the lake is facilitated by weak expression of surface drain processes. They are interfered with the high density of vegetation cover, which is formed in the absence of anthropogenic load. According to the degree of contamination, ecosystems of the watershed area may be arranged in the following row: steppe fescue meadow> secondary brome post-forest meadow > border birch forest > birch forest> ruderal meadow. Contribution of contemporary gas-aerosol emissions by PA "Mayak" in contamination of the studied territory was evaluated according to the level of radionuclide content in forest litter. Within the soil profile the forest litter was marked as the indicating layer, which summarizes content of radionuclide pollutants of vegetation cover, introduced via root and aerial ways. Investigations were carried out in forest areas of western and eastern transects, identical in composition of plant communities and soil and ecological conditions. They were located almost at the same distance from PA "Mayak". Since current gas-aerosol emissions of PA "Mayak" mainly include ^{137}Cs among long-lived radionuclides, while ^{239,240}Pu is practically absent [9], below is the represented data on ^{90}Sr and ^{137}Cs (Table 2).

Table 1. The radionuclide stocks in the soil of the different ecosystem types, kBq/m²

Type of geochemical interfaces	^{90}Sr	^{137}Cs	^{239,240}Pu	^{90}Sr / ^{137}Cs	^{90}Sr / ^{239,240}Pu
Coastal area of the lake					
water edge, reeds	270±50	60±10	1.3±0.2	4.6	207
beach zone, reed-sedge brakes	2 300±630	210±50	7.3±0.9	11.3	315
Watershed area					
birch forest	8 150±200	310±90	25.0±6.0	26.2	326
border of the birch forest	10 600±350	410±100	26.4±6.4	25.7	401
steppe fescue meadow	16 700±3 540	700±200	62.2±15	23.8	268
secondary brome post-forest meadow (native)	13 900±2 900	400±100	33.2±9.5	34.7	418
ruderal meadow (anthropogenic disturbed)	6 700±270	200±60	15.9±1.6	35.9	421

Table 2. The radionuclides content in the forest litter and soil

Samples	Transects	^{90}Sr, Bq/m²	^{137}Cs, Bq/m²	^{90}Sr/^{137}Cs
Litter	Western	500±370	20±5	25.0±10.9
	Eastern	1 200±300	500±100	2.4±0.8
Soil, 0-15 cm	Western	18 600±4 500	8 900±3 400	2.1±0.4
	Eastern	58 300±13 300	81 700±21 800	0.7±0.2

Table 3. The radioactive inventories in soils of EURT (in brackets confidence intervals are given), n×10^{12} Bq

Locality	^{90}Sr	^{137}Cs	^{239,240}Pu
Impact	428.2 (286.4–643.4)	15.1 (10.2–22.4)	1.175 (0.623–2.242)
Buffer	143.0 (70.0–357.0)	51.0 (26.0–56.0)	0.5 (0.4–0.6)
Total	571.2	66.1	1.675

The average content of ^{90}Sr at forest litter of the western transect makes 500 Bq/m^2, and the eastern transect is 2.5 times more. The same refers to difference in levels of soil contamination with ^{90}Sr on these areas. Content of ^{137}Cs in forest litter of the eastern transect is 25 times greater than that of the western one, while the level of soil contamination differs only 9 times. Relative enrichment of forest litter with ^{137}Cs is also indicated by the change in ratio ^{90}Sr/^{137}Cs. Increasing content of ^{137}Cs in forest litter of the eastern transect, not proportional to the increase in soil contamination level, might have been be caused by the additional inflow consisting of gaseous and aerosol emissions from PA "Mayak". As noted above, EURT territory was exposed to additional contamination. In 1967, a winter with little snow was followed by an extremely dry summer. Karachay Lake, which was used by PA "Mayak" as an open repository of radioactive waste, became very shallow. Bottom sediments contaminated mostly with ^{137}Cs, had been rising with the wind and carried the long distances. We made a quantitative evaluation of the contributions of these two events as to contamination of soil and vegetation cover of EURT on the impact and its adjacent areas of the buffer zone of the trace. Calculations were made with the use of values of radionuclide ratios that characterize precipitations of various origins, and stocks of radionuclides at reference sites. We also took into consideration the level of global background and contribution of gaseous and aerosol emissions from PA "Mayak". Previously it was found that in the accident fallout in 1957 the ratio value of ^{90}Sr/^{137}Cs was 71.0, while it was 0.3 in the contaminated bottoms sediments of Karachay Lake shores. Solving the system of equations using the above ratios, we calculated the contribution of the two accidents into contamination with ^{90}Sr and ^{137}Cs of soil cover at each transect:

$$x+y=a, \qquad (1)$$

$$v+p=b, \tag{2}$$

$$-71x+v=0, \tag{3}$$

$$-0.3y+p=0, \tag{4}$$

where x – ^{137}Cs, v – ^{90}Sr, received during the accident in 1957 (kBq/m^2);

y – ^{137}Cs, p – ^{90}Sr, brought with the lake bottom sediments of Karachay Lake (kBq/m^2);

a – total content of ^{137}Cs, b – ^{90}Sr in the sampling areas (kBq/m^2). More detailed calculations of the contribution of various sources into radioactive contamination of soils are given [10]. As one would expect, the main contribution (97%) in the contamination of the central transect was made by the Kyshtym accident, on the western one it constituted 67% and on the eastern one – 31%, based on the total content of radionuclides in soils. Wind transfer of sediments of Karachay Lake added to this contamination 3%, 26% and 63% respectively. The contribution of global fallout and emissions of PA "Mayak" into pollution of the central transect is very small if compared with emergency inflow (<0.1%) and in adjacent territories it does not exceed 7%. Data characterizing the levels of contamination of soil and vegetation cover and patterns of spatial distribution of radionuclides allowed us to assess their integral stock on the study area (Table 3). It is clear that the soil of the impact zone have a concentration of the principal amount of ^{90}Sr (75%) and plutonium isotopes (70%). At the same time most of the ^{137}Cs (77%) is located in the buffer area. This imbalance in the distribution of radionuclides is conditioned by the fact that contamination of the buffer area was mostly contributed by the Karachay accident in 1967.

1.2. Features of Uptake of Long-Lived Radionuclides by Plants

It is known that the uptake of radionuclides by plants depends on their specific features, as well as the complex of soil and climate and environmental factors. During study of the accumulation ability of plants on EURT territory we took into account the wide diversity of plant communities, significant trace length and change of density of soil contamination with the distance. Therefore, samples of herbs and certain species of herbaceous plants, commonly grown at EURT, have been selected on a limited area in accordance with their ranking within the impact and buffer zones.

In the first stages of research we assessed the intrapopulation variation of the accumulation ability of plants as to ^{90}Sr, ^{137}Cs and ^{239,240}Pu. In addition we made special sampling of some species (5 replications for each species) in a limited area of EURT with an average density of soil contamination with ^{90}Sr is 13200 kBq/m^2, ^{137}Cs – 500 kBq/m^2 and ^{239,240}Pu – 50 kBq/m^2. The level of soil contamination with radionuclides is reflected in their content in the aboveground plant mass (Table 4).

Table 4. The specific activity of radionuclides in the aboveground mass of plants growing on limited EURT area, Bq/kg of the air-dry matter

Species	^{90}Sr		^{137}Cs		^{239,240}Pu	
	$\bar{x} \pm m_{\bar{x}}$ (min-max)	CV, %	$\bar{x} \pm m_{\bar{x}}$ (min-max)	CV, %	$\bar{x} \pm m_{\bar{x}}$ (min-max)	CV, %
Trifolium medium L.	419 260±78 025 (228 000-591 600)	37.2	304.8±72.3 (193-530)	47.4	3.23±0.79 (0.56-5.76)	58.8
Lathyrus pratensis L.	37 888±17 595 (3 504-85 496)	92.9	133.8±78.8 (23.0-390.5)	117.8	0.13±0.06 (0.06-0.13)	77.7
Urtica dioica L.	71 876±18 387 (21 360-107 044)	51.2	22.2±6.9 (6.9-39.1)	62.2	0.78±0.37 (0.14-1.78)	82.4
Leonurus quinquelobatus Gilib.	23 060±4 730 (9 038-31 080)	41.0	10.0±2.4 (5.8-17.2)	48.0	0.32±0.14 (0.02-0.69)	76.9
Arctium tomentosum Mill.	8 114±3 741 (3 907-21 341)	92.2	14.0±4.2 (8.4-28.2)	60.0	0.52±0.22 (0.13-1.16)	67.0
Rumex confertus Willd.	7 028±2 966 (1 056-14 877)	84.4	6.4±1.2 (3.9-9.3)	37.5	0.30±0.09 (0.01-0.64)	36.6
Bromopsis inermis (Leys). Holub	5 528±1 339 (1 596-7 838)	48.4	5.7±1.0 (3.1-7.7)	35.1	0.58±0.19 (0.27-1.2)	62.5

Thus maximum specific activity in the aboveground mass of the species studied was noted for ^{90}Sr. Content of ^{137}Cs varies within 5–304 Bq/kg, and ^{239,240}Pu within 0.13–3.23 Bq/kg. A wide range of specific activity of each of the radionuclides in the aboveground mass of plants grown under identical conditions is conditioned by the species specificity. Attention should be paid to the high content of ^{90}Sr in *Trifolium medium* and *Urtica dioica*, and the lowest in *Bromopsis inermis*. The highest specific activity of ^{137}Cs (130 and

305 Bq/kg) was found in the representatives of the *Liguminosae* family, and $^{239,240}Pu$ in *Trifolium medium*. High volatility in accumulation ability within a population of one plant species is marked for ^{90}Sr and $^{239,240}Pu$. So the specific activity of these radionuclides in the aboveground mass of one plant species may vary within the order of magnitude, and coefficients of variation (*CV*) may reach 82–93%.

High intrapopulation diversity of one species of plants according to accumulation of ^{90}Sr was also observed in the work [11]. Intrapopulation differences between specimens in their ability to accumulate ^{137}Cs appeared to a lesser extent. Its specific activity varies, usually 2–3 times under the coefficient of variation 35–62%. An exception is *Lathyrus pratensis*, with a quantity of 117.8%.

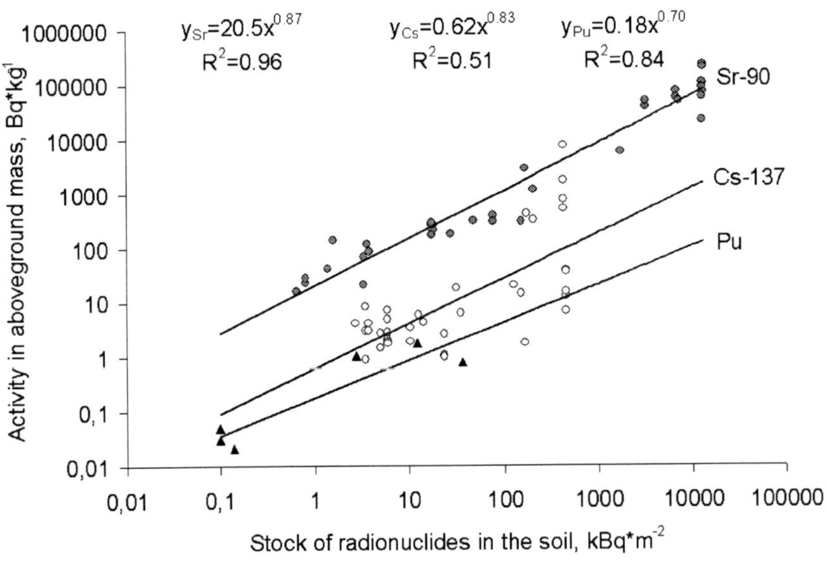

Figure 3. The specific activity in the *Urtica dioica* L. aboveground mass depending on stock of radionuclides in the soils.

A wide range of pollution levels allowed us to estimate regularities of inflow of the main pollutants (^{90}Sr and ^{137}Cs) into the plants, depending on their stocks in the soils. By the examples of common species *Urtica dioica* L. we showed that the content of radionuclides in it does not vary in direct proportion to the level of soil contamination, but is reliably described by power function with a coefficient of determination $R^2 = 0.96$ for ^{90}Sr, 0.51 for

^{137}Cs and 0.84 for Pu respectively (Figure 3). Similar dependency was observed for other species and, in general, for herbs.

Table 5. The transition coefficients (m^2 $кг^{-1}$$\times 10^{-3}$) of radionuclides to plants

Species	^{90}Sr			^{137}Cs		
	Impact	Buffer	Background	Impact	Buffer	Background
Melandrium album (Mill.)	8.8±2.0	10.1	55.2	0.12±0.04	1.42±0.13	1.11
Stellaria media (L.) Cyr.	4.0	25.6	60.0±32.6	0.05	0.47	0.72±0.07
Urtica dioica L.	8.7±1.6	13.0±1.8	29.7±2.9	0.45±0.25	0.60±0.17	0.60±0.15
Rumex confertus Willd.	1.9±0.6	8.7±2.4	9.4±2.9	0.03±0.01	0.79±0.27	0.60±0.20
Arctium tomentosum Mill.	1.2±0.5	3.9±0.4	15.8±4.6	0.17±0.11	0.09±0.04	0.37±0.12
Cirsium setosum (Willd.) Bess.	7.7±1.7	–	31.5±12.4	0.30±0.13	–	–
Achillea millefolium L.	5.1±3.5	3.0±0.5	14.9±6.9	0.95±0.31	0.16±0.02	2.19±1.20
Tanacetum vulgare L.	7.6	9.6	37.0±13.0	0.31	0.17	0.15±0.03
Artemisia vulgaris L.	5.0 ±1.1	3.0±2.7	6.8±3.8	0.06±0.01	0.16±0.06	0.14±0.02
Leonurus quinquelobatus Gilib.	1.2±0.2	3.7±1.8	18.6±2.7	0.016±0.004	0.10±0.00	0.09±0.01
Pulmonaria mollissima M. Pop.	25.2±1.8	–	38.0	0.91±0.06	–	1.9±0.2
Plantago major L.	10.4±3.1	6.7	55.6±60.5	2.2±1.3	0.6	5.9±1.2
Mean±S.D.	7.2±2.4	8.7±2.5	33.5±6.1	0.46±0.24	0.46±0.13	1.35±0.58

Conditioned by a wide range of soil contamination it is reasonable to characterize the accumulation ability of plants according to the most informative indicator – transition coefficient ($m^2/kg \cdot 10^{-3}$). In the surveyed areas all examined species were characterized with a high variability of these coefficients, depending on the specific features and conditions for plant growth (Table 5). However, general rules that determine the uptake of radionuclides to plants showed up clearly. Mathematical analysis has shown that plants growing in the impact zone usually accumulate radionuclides to a lesser extent than in the background territory. Comparison of these samples by Student's test showed highly reliable differences (for ^{90}Sr $t_{exp} = 44.2$; for ^{137}Cs $t_{exp} = 8.6$ at $t_{tabl} = 2.83$ and $P = 0.05$). It can be assumed that contributingto this decrease of accumulating ability of plants is the specific physical-chemical state of radionuclides in soils, as well as barrier-regulating mechanisms, which are formed in the process of plant adaptation to conditions of chronic pollution of the environment. In all cases, the accumulating ability of plants as to ^{90}Sr is 1–2 degrees greater than that for ^{137}Cs.

1.3. Changes of the Radiation Dose from the Time of the Accident to the Present

Since the Kyshtym accident in 1957 the radioecological situation in the area of the East-Ural Radioactive Trace has undergone significant changes. Scattered at the explosion radioactive waste contained mainly short-lived radionuclides, the most significant contribution was made by $^{144}Ce + ^{144}Pr$ and $^{95}Zr + ^{95}Nb$. The main danger in the long term was long-lived ^{90}Sr, kept in balance with the daughter product of decay – ^{90}Y: their content in the isotopic mixture was about 5.4%. Laid-down mixture was characterized by the presence of gamma-radiation with a total energy at the time of trace formation 7.63 MeV per decay of ^{90}Sr, accepted as "the frame" radionuclide, as well as beta-radiation with a total activity approximately 3 times greater than the gamma-rays component. Living objects in EURT territory received the main part of the radiation dose during the first four years after the accident. Over the years, the energy of gamma radiation decreased from 7.6 to 0.004 MeV per decay of ^{90}Sr, which resulted in the ninetieth years in reduction of the exposure dose of gamma-radiation at 1 m height in 2800 times [12].

In our study the contribution of the external γ-radiation from natural and artificial radionuclides was calculated based on measurements made with a dosimeter-radiometer MKS-AT 1117MBDPB-01 and dosimeter DRG-01T. At the background territory the exposure dose rate (gamma-rays) was 9−11 μR/h, at the buffer zone − 15−20 μR/h, and at the impact zone it ranged within 20−157 μR/h. During evaluation of the doses accumulated by plants due to ^{90}Sr and ^{137}Cs, we used several approaches. First, on the basis of concentrations of radionuclides in topsoil we calculated absorbed doses to the organs of plants, located in 0–5 cm soil layer. This approach was especially important for cryptophytes and hemicryptophytes plants that have renewal gemmae located on the soil surface or buried into the topsoil. The absorbed dose rate (R) from incorporated radionuclides was calculated according to the model [13]: $R = q_1 L_{(Sr+Y)} + q_2 L_{(Cs)}$, where q_1 and q_2 – measured specific activity of ^{90}Sr and ^{137}Cs in the soil or vegetative mass; L – absorbed dose rate (cGy/s) created by this radionuclide inside a uniformly contaminated volume at $q_0 = 3.7\times10^4$ Bq/g. This simple model allows us to estimate, at the first approximation, the absorbed doses of different species of plants and compare cenopopulations from different EURT zones according to this parameter. As an example, Table 6 shows the results of calculation of accumulated doses for cryptophytes growing in the impact, buffer and background zones.

It is clear that the absorbed doses calculated for plants in the buffer areas, are usually greater by one order of magnitude than the background areas, and in impact coenopopulations differences with background plantings make 3-4 orders of magnitude. If one takes into account that the lifetime of many herbaceous plants can reach 20 years, plagiotropic parts of cryptophytes can accumulate doses of about 7–9 Gy during the entire period of ontogeny.

Table 6. The specific activity of radionuclides in 5 cm soil layer and caused by them radiation doses for renewal gemmaes of cryptophytes

Zone	Community type	Specific activity, Bq/kg		Exposure dose rate, C/kg×h	Absorbed dose rate, mGy/year
		^{90}Sr	^{137}Cs		
Impact	Dry secondary forbs-grass meadow	113 135–122 750	5 141–5 160	$9.8–11\times10^{-7}$	334.0–362.2
	Border of the birch forest	141 550	7 040	1.2×10^{-6}	419.7
Buffer	Birch outlier	646	314	8.2×10^{-9}	2.8
Background	Old-ploughed field	15	31	4.1×10^{-10}	0.14

Table 7. The absorbed dose changes depending on the distance from the soil surface

Investigated plot	Distance from the surface soil, cm	Absorbed dose rate from ^{90}Sr+^{90}Y, mGy/month
Forest border, N 55°46.47 E 60°53.11	0	3.9
	10	2.5
	20	2.3
	30	2.1
	40	2.0
	50	1.8
	60	1.7
	70	1.6
	80	1.5
	90	1.4
	100	1.3

Obviously these doses refer to small doses for vegetable objects. In order to estimate the radiation dose on the leaves and stems of plants we used coefficients characterizing attenuation of the radiant flux depending on the distance from the soil surface (Table 7). We also took into account irradiation of plants by radionuclides accumulated in the stems and leaves. Thereto, we measured the power flux in the grass, then the aboveground mass of plants at the site was cut and measurements were repeated. The results of calculations of the absorbed dose with a glance to radionuclides accumulated in the vegetative mass of plants showed that in areas with dominating *Poaceae* family (*Bromopsis inermis*), the contribution of incorporated radionuclides is small (from 1.6 to 4.6%). The main source of irradiation is the soil that accumulated a maximum of radionuclides. On sites with dominating *Urtica dioica* – accumulator of ^{90}Sr, the situation is different, and contribution of radionuclides incorporated in vegetative organs varies from 2.8% in the ground surface (1–10 cm) to 28.5% at the height of 80–100 cm.

1.4. Brief Analysis of the State of Flora and Vegetation in the EURT Area

The situation in the impact zone of EURT ruled out the possibility of human habitation, therefore this territory became a unique testing area for scientific research to assess the state of vegetation for the long-term influence of ionizing radiation. First open publications on the assessment of the Kyshtym accident showed that substantial changes in composition and structure of phytocoenoses in the EURT head area appeared within 2 years after the accident. Coniferous trees that first died received doses over 30–40 Gy during the autumn-winter period. Lethal doses, which caused the death of birch stands, were about 5 times greater. The greatest damage was observed in the lower and middle part of the crown. The top level was blown by the wind and washed by rain so it was damaged to a lesser extent. In some cases, changes in community composition were caused not by the direct effect of radiation, but indirect factors. So, 2 years after the accident the defoliation of trees, and partial drying and destruction of trunks occurred in strongly damaged forests. The increase in brightness, temperature and precipitation contributed to a better development of light-demanding herbaceous plants [14]. At plots where the pines perished and birch trees were damaged, the phytomass of grasses increased by 3–5 times, mainly due to reproduction of *Carex*, *Trifolium*, *Calamagrostis*, *Poa* and others. Multiple morphoses were observed in the newly emerging leaves and shoots, though abnormalities were temporary; 3-4 years later the most irradiated of the trees did not differ from the control specimens. Phenological observations showed that foliage expansion and further blossoming of the trees on the contaminated areas was delayed.

From the structure of herbaceous communities, cryptophytes and hemicryptophytes, that have renewal buds located in soil or on the surface, have fallen out completely or reduced their projective cover during the first years. Therophytes reproduced only by seeds have fallen out completely [15].

Similar results were obtained in the course of studies performed in the U.S. at old fields near nuclear weapons testing sites [16]. Additionally, large-scale experiments with irradiation of natural communities by powerful γ-sources, held in the tropical forests of Puerto Rico, broad-leaved forests in Wisconsin, Mediterranean-type forests in France, and coniferous-deciduous forests in the state of Georgia and in Central Russia [17-20] showed that plant communities can survive a fairly intense irradiation. Significant rearrangements in them are observed at cumulative doses over 25 Gy.

Since the date of the Kyshtym accident the radiation doses significantly decreased, and succession processes in phytocoenoses largely regained their former species composition. It is of interest to assess vegetation at EURT territory 50 years after the accident.

Our investigations show that the EURT vegetation presently comprises a wide variety of ecosystem types [7] and high species diversity. A detailed description of the major types from the most contaminated (impact) zone is listed below. Currently, the vegetation here is typical for the steppe zone and is characteristic for the region. Birch forests, grassy steppe communities, as well as small groves of birch and aspen, located in the relief lowering, and rarely pine forests, are represented here. Prior to the accident this area included a small village, arable land and meadows. After 1957, the fields were abandoned, forming temporary plant communities, which represent stages in the process of restoring natural vegetation [15]. On the place of evicted villages recultivation works were carried out: buildings were demolished, the soil plowed to a depth of 70 cm or the upper and most polluted layer was removed. Such areas became overgrown with the nitrophilous vegetation usual for this area. It is still growing because the clay soil firmly retains nitrogen. 50 years after the accident, most of the fallows passed the successional stage, which precedes either the formation of a young forest, or steppe, depending on soil type, i.e. stage of initial phytocoenoses.

Forest communities occupy up to 45% of the head of EURT. In most cases, the age of the trees is about 60–70 years. They are represented by herbaceous birch forests (forb birch forest, forb-reedgrass pure birchwood and forb-reedgrass birch forest with a single pine and aspen, willow-herb-reedgrass birch forest, sedge-grass-brome-reedgrass birch forest, stone bramble-sedge grass birch forest, steppe-meadow-seseli birch forest).

Total projective cover (PC) of these forests is usually up to 70%, sometimes 90%. Forests stand is represented with the *Betula pendula*, with some *Pinus silvestris* and *Populus tremula*. In the second sublayer one can find several species of *Salix* and *Padus*, *Sorbus aucuparia* and *Acer negundo*. Undergrowth is formed by the same forest-forming species. PC of undergrowth is rather small, only 5–10%. The density of the canopy is usually 40–60%, rarely 70%. The composition of shrub layer includes *Chamaecytisus ruthenicus*, *Genista tinctoria, Cerasus fruticosa*, etc.

Dominants of the herbaceous plants of forests are usually *Calamagrostis arundinacea, Rubus saxatilis.* In a steppe version of a birch forest the dominant species include *Festuca rubra, Seseli libanotis, Carex praecox.* Accompanying species with high degree of constancy are grasses, sedges, as well as *Cirsium setosum*, *Galium boreale, Fragaria vesca*, etc. Certain species growing here (*Galium verum, Seseli libanotis, Trifolium montanum*, etc.) indicate that the reedgrass birch forests bear signs of steppe as well. Besides that, ruderal and meadow-ruderal plants can be found in the forests quite often. The height of the grass stand varies from 50 to 100 cm. Species diversity ranges from 29 to 65 species. Moss cover is almost unexpressed.

Special attention should be given to the most transformed forest community in the area of the central axis in 6 km from the epicenter of the accident – ruderal lungwort-thistle community, developed on the place of forb-reedgrass birch forest. Total PC is 100%. There are mature birch trees; the undergrowth is formed by small numbers of aspen and pine trees. The bushes are represented only by *Rosa majalis*. The height of the grass stand reaches 100–120 cm. The dominants here are *Cirsium setosum* and *Pulmonaria mollissima*, codominants are presented by *Lathyrus pratensis*. Besides that, forest, meadow and forest-meadow and ruderal species grow here (33 species registered in total).

Old-growth brome- willow-herb aspen forest grows on the western periphery of trace. Density of the forest stand is 40–50%. Undergrowth is formed by a large number of *Populus tremula* with rare *Prunus padus* and *Betula pubescens*). Shrub synfolium is made up mainly of the *Rosa canina.* The composition of herbaceous synfolium consists of more than 35 meadow-forest species with 90% of PC; height of the grass stand is 100–120 cm. Dominating species are *Epilobium angustifolium* and *Bromopsis inermis.*

The Karabolka River flows along the central axis of the impact zone (see Figure 1). Tall bushes of *Salix* and *Carex* grow along its banks in a narrow strip. Total PC reaches 70%. Undergrowth is formed by single young birch trees. Shrub synfolium is formed mainly by *Salix pentandra* and *S. caprea* with rare *Ribes nigrum*. The height of the grass stand varies between 70 and 90 cm. About 30 species grow in the community. Dominants in this phytocenosis are *Carex rostrata* and *Filipendula ulmaria*. Associated species, as well as the dominants, are typical for moistened habitats; *Equisetum fluviatile, Scirpus sylvaticus, Lycopus europaeus* were found here.

Herbaceous communities occupy about 40% of the impact zone. Grassland coenoses mostly represent post-forest meadows (secondary cow-parsnip-brome meadow with raspberries, secondary forb-brome meadow, secondary forb-grass meadow with stone bramble, secondary forb-thistle-brome meadow with birch). High PC – up to 100% - is typical for such plant communities.

By the end of vegetation period grass stand reaches 80-120 cm. There are single species of young trees (*Betula pendula*) and bushes (*Spiraea crenata, Rosa majalis, Ribes nigrum*). Meadow-ruderal, meadow and meadow-forest species, such as *Bromopsis inermis, Calamagrostis epigeios, Poa pratensis,* come to the fore recently. In most cases they are associated with usual meadow species – *Dactylis glomerata, Achillea millefolium, Knautia arvensis,* different species of *Vicia, Trifolium, Lathyrus, Veronica*, as well as forest species – *Fragaria vesca, Chamaenerion angustifolium, Geranium sylvaticum.* Some ruderal species were found in the studied communities – *Cirsium setosum, Sonchus arvensis, Urtica dioica, Lithospermum officinale*, though they are not abundant.

In phytocoenoses of steppe areas and meadows on the shores of the Berdenish and Uruskul Lakes (see Figure 1) vegetation has largely retained its original appearance. Such is the steppe fescue meadow adjacent to the low, relatively gentle southern slopes of the limestone outcrops near Berdenish Lake. Total PC reaches 100%. Species diversity reached 65 species. There are single young birch trees, shrub layer is more expressed and is formed mainly by *Spiraea crenata*. Grasses and sedges (*Festuca pseudovina, Stipa capillata, Carex praecox*) dominate in the grass layer. Typical steppe meadow species such as *Galium verum, Artemisia commutata, Silene baschkirorum* are frequently met. However, ruderal species as *Cirsium setosum, Berteroa incana, Leonurus quinquelobatus* also penetrate into these communities.

The *Phragmites communis* brakes with *Salis* are adjacent to the shores of lakes and the edges of marshes in the northern part of the impact zone. The total PC in these phytocoenoses reaches 50%. Shrub layer forms *Salix pentandra* with a fairly high abundance – up to 15%. The birch grows occasionally. The height of the grass stand is very high and reaches 90-150 cm. The high degree of moisture probably has a restricting influence, and therefore species diversity here is extremely poor - only 9 species. The *Phragmites* dominate in this cenosis, the other species are met in single specimens (*Cirsium setosum, Lysimachia vulgaris, Solanum dulcamara*).

A summary of the above-listed data leads us to the conclusion that natural ecosystems practically went off the recovery stages. Its status, diversity and species resources have a very high estimation. Those anthropogenically transformed in the EURT zone localities deserve special attention. Our investigations show that recovery successions go very slowly. The relationship between radiation and successions is not detected. The land anthropogenic use is a defining factor of the disturbance level of these communities.

Anthropogenically transformed communities (ruderal thistle-brome community, meadow-ruderal thistle-bluegrass-brome community, ruderal beakchervil-thistle-nettle community) occupy areas on the southern shore of Berdenish Lake (see Figure 1). Prior to the accident these areas sheltered a village, subsequently subjected to recultivation. Total PC is 80-90%. The height of the grass stand varies from 50 to 150 cm. Species diversity is relatively low (12 – 32 species). These communities are characterized by a high proportion of ruderal plants with such species as *Bromopsis inermis*, *Cirsium setosum* and *Urtica dioica* dominating in the community, forming extensive brakes and suppressing other species. Besides that, they form a rough, recalcitrant ground litter, which also prevents germination of less aggressive species, and, as a result, a further regenerative succession of communities. *Anthriscus sylvestris, Elytrigia repens, Melandryum album* are found quite often.

In work [21] E. Smirnov gave a detailed analysis of flora in the head part of EURT taking into account life forms, distinguished by anatomical and morphological and ecological features. The influence of life forms of plants and their ability to accumulate radionuclides reveals itself only when it is peculiarly adapted to the specific conditions of growth (high humidity or salinity) or has a distinctive root system structure. Radio-sensitivity of the species largely depends on the life forms of plants, in particular, on location of the renewal buds with reference to the source of ionizing radiation. E.G. Smirnov identified several species that have increased in number on the fourth year after the accident, when short-lived radionuclides decayed and the radiation dose was decreased by two orders of magnitude. Among them – *Cirsium setosum, Taraxacum officinale, Chamaenerion angustifolium, Sonchus arvensis, Potentilla anserine, Poa pratensis.* It was assumed that these species would occupy a dominant position on the EURT territory.

Our research showed that the forecast proved to be correct as to *Cirsium setosum*. This species takes a special place in the community of EURT, it is found in most of the described communities. It is significant to note that at the most contaminated site located 6 km from the epicenter of the accident, the *Cirsium setosum* population has over 70% of plants with albino-type chlorophyll mutation (Figure 4). Growing plants in experimental conditions from seeds collected in this cenopopulation confirmed that this mutation is inherited by next generations.

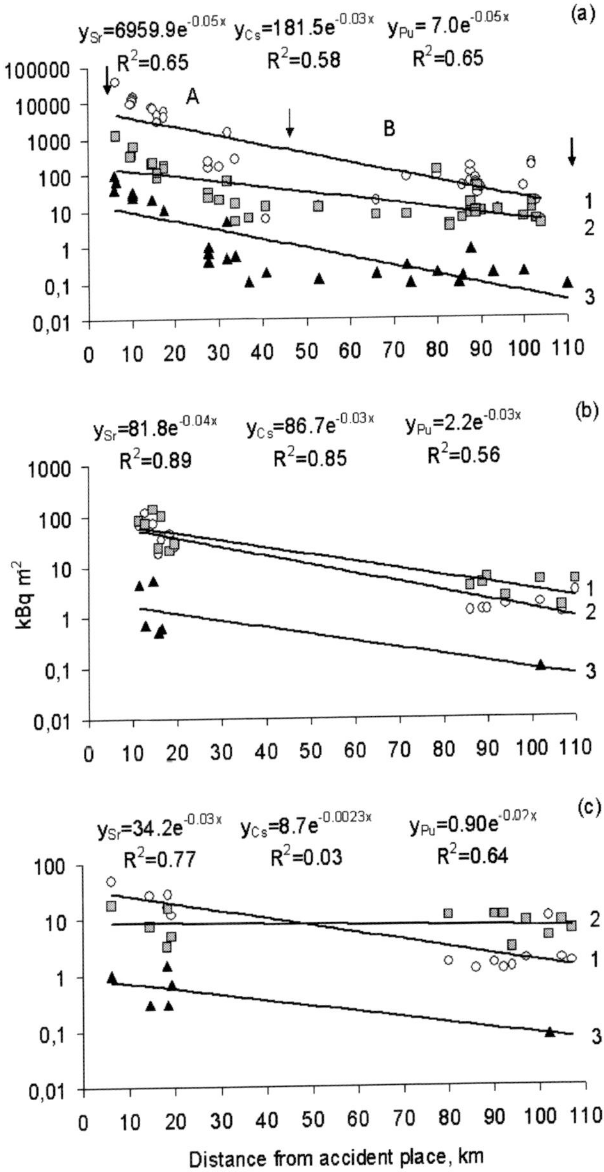

Figure 4. Stock of rdionuclides in soils of the impact (A) and buffer (B) zones of the.EURT. The transects: *a* – central, *b* – eastern, *c* – western. 1 – ^{90}Sr, 2 – ^{137}Cs, 3 – $^{239, 240}$Pu.

It can be assumed that due to weakening competition from more radiosensitive plants, aggressive *Cirsium setosum*, which is a vegetative mobile with numerous seeds, has a distinct advantage and scattered throughout EURT territory, appearing not only in synanthropic phytocoenoses, but also in quasinatural communities.

We also discovered that annual and biennial plants, reproducing solely by seeds, are rarely found at the studied EURT sites – mostly in phytocoenoses of the little-used forest roads. Perhaps they are regularly carried here with a transport.

A comparison of impact zone geobotanical descriptions with their adjacent buffer territory (group of Tygish, Chervyanoe, Sungul Lakes, and see Figure 1) showed that the species composition of plant communities is similar to that of the coastal edges of the Berdenish and Uruskul Lakes. However, most of the buffer zone is a residential area. Many villages are located in this territory. Therefore, it is rather hard to maintain the direct comparison of buffer and impact zones. In most cases aligned watersheds and gentle slopes are used for tillage and pasture. Grazing lands are generally occupied with natural meadow vegetation with interplanting grain crops and legumes. A significant part of the investigated area of the buffer zone is covered with mixed birch and pine forests and forest outliers. Shrub synfolium of such forests have *Rubus idaeus, Rosa majalis, Ribes nigrum*. Herb synfolium is represented by *Rubus saxatilis, Fragaria vesca, Glechoma hederacea, Trifolium pratense*, abundant *Geum rivale, G. aleppicum*.

Summarizing the foregoing material, we can say that the vegetation of EURT now presents a complex of synanthropic and semi-natural plant communities at different stages of restoration successions. The steppe nature of the studied forest communities can be attributed to increase of brightness and heating of the soil due to dilution of the forest canopy that has arisen under the influence of radiation.

In this case we may state a development of the secondary effects of irradiation. The involvement of ruderal species in meadows and forest coenoses of the impact zone is reduced, while plants typical to the communities play a significant role. Species diversity is very high, which is associated with low anthropogenic load.

It is significant to note that the current state of phytocoenoses in place at the demolished local villages is still largely determined by the degree of man's impact in the pre-accident period and recultivation activities.

In recent studies by other authors [22], species diversity and the state of plant communities at the head of EURT is evaluated very high. This area preserves 16 species of plants listed in the “Red Book”. Despite the high levels of radioactive contamination, this zone is regarded as the core of ecological frame of the forest-steppe zone, which plays an important role in preserving biodiversity and maintaining ecological stability in the region.

Having agreed with such high assessment of the diversity of ecosystems in this zone and their species richness, we believe that the quality of the gene pool of flora can be detected only in the course of special ecological and genetic research. The presence of genetic damage in plants of the EURT area is indicated by a high frequency of chromosomal aberrations and rare alleles of certain enzymes [23-28], increased variability in vital capacity and radioresistance of contemporary brood, and a large number of different morphoses detected at the early stages of ontogenesis [5,29]. This theme is very wide and requires an independent review.

CONCLUSION

The Kyshtym accident in 1957 at the nuclear facility of "Mayak" (Southern Urals) has led to the formation of the East-Ural radioactive trace. Taking into account the length of the investigated area, it was ranked by the level of radionuclide contamination with selection of the impact and buffer zones. The study of the spatial distribution of radionuclides in the trace showed that within the trace axis and its eastern periphery the changes in their stocks were satisfactorily approximated by a power function with increasing distance from the epicenter of the accident. At the same time, radionuclides are arranged in a row $^{90}Sr \geq {}^{137}Cs > {}^{239,240}Pu$ according to the level of ground contamination. Content of ^{137}Cs in soil on the western periphery of the trace does not depend on the distance. Study of geochemical integrations, including the coastal zones of aquatic ecosystems, water-logged grounds and watershed areas revealed contribution of ecosystem types, landscape diversity and recultivation activities in the intensity and direction of migration flows of radionuclides. We noted that decontamination processes dominate at the depressed elements of the landscape.

The complex nature of soil contamination in the investigated area is conditioned by contribution of another accident – the wind transfer of contaminated bottom sediments from the shores of Karachay Lake being an open storage of liquid radioactive waste by PA "Mayak". ^{137}Cs was the main contaminant. Certain influence is exerted by contemporary atmospheric emissions by PA "Mayak". Currently, the impact zone is mostly (95.7%) contaminated with ^{90}Sr after emergency emissions in 1957. The western periphery is distinguished by a relative decrease of stocks of ^{90}Sr and growth of ^{137}Cs ones after the Karachay accident. Contamination of the eastern part of the trace is formed mostly (63%) by the same source.

With a wide range of radionuclide contamination of soils, specific activity of ^{90}Sr, ^{137}Cs and $^{239,240}Pu$ in certain species of plants in general vary in accordance with a power function. The values of transition coefficients that characterize accumulative ability of the plants vary within wide limits. There is a clearly expressed reduction of accumulative ability of the plants growing in the impact zone, and a high intrapopulation diversity of plants according to their ability to accumulate the radionuclides.

Living objects in EURT territory received the main part of the radiation dose during first four years after the accident. Over the years, the energy of gamma-radiation decreased from 7.6 to 0.004 MeV per decay of ^{90}Sr. Our study showed that the absorbed doses calculated for plants in the buffer areas are usually greater by one order of magnitude than in the background areas, and in impact coenopopulations differences with background plantings are 3–4 orders of magnitude. If one takes into account that the lifetime of many herbaceous plants can reach 20 years, plagiotropic parts of cryptophytes can accumulate doses of about 7–9 Gy during the entire period of ontogeny. Obviously these doses refer to small doses for plants.

Vegetation of EURT now presents a complex of synanthropic and semi-natural plant communities at different stages of restoration successions. Forest communities have signs of steppe nature that can be attributed to increased brightness and heating of the soil due to the open forest canopy that has arisen in the first years under the influence of radiation. These changes indicate secondary effects of irradiation in ecosystems. The involvement of ruderal species in all coenoses of the impact zone is reduced, while plants typical to forests and meadows play a significant role. Species diversity in the impact zone is very high; some communities have up to 65 species, which is associated with a low anthropogenic load. It is significant to note that the current state of phytocoenoses on the land of the demolished local villages is still largely determined by the degree of man's impact in the pre-accident period and recultivation activities. The natural ecosystems successfully recovered at a faster pace.

ACKNOWLEDGMENTS

This study was completed with support by the Russian Foundation for Basic Research (project no. 11-04-01260), the Interdisciplinary Projects of the

Ural Division of the Russian Academy of Sciences (no. 09-M-24-2001) and the Program for Directional and Fundamental Investigations of the Ural Division of the Russian Academy of Sciences (DFI-1, №11-4-01-NC).

References

[1] P.L. Gorchakovsky, "The vegetation cover of the Ural", Moscow (1968) pp111-262. (in Russian).

[2] F.I. Pavlotskaya, G.A. Fedoseev, E.V. Babicheva, L.N. Zatsepina and E.B. Tiuriukanova, *Pochvovedenie*, N2 (1964). (in Russian).

[3] Q. Chen, A. Aarkrog, S.P. Nielsen, H. Dahlgaard, H. Nies, Y. Yu and K. Mandrup, *J. of Radioanalytical and Nuclear Chemistry*, 172, N2 (1993).

[4] Field geobotany, Moscow, Leningrad (1964) (in Russian).

[5] V.N. Pozolotina, I.V. Molchanova, E.N. Karavaeva, L.N. Mikhaylovskaya, and E.V. Antonova, "The current state of terrestrial ecosystems in the Eastern Ural Radioactive trace: contamination levels and biological effects", Yekaterinburg (2008). (in Russian).

[6] I.V. Molchanova, E.N. Karavaeva, and L.N. Mikhaylovskaya, "Radioecological investigation of soil-plant cover", Yekaterinburg (2006). (in Russian).

[7] V.P. Firsova, I.V. Molchanova, P.V. Meshcheryakov, T.S. Pavlova, E.N. Karavaeva, E.V. Prokopovich and V.V. Toshchev, "Soil and ecological conditions of accumulation and redistribution of radionuclides in the zone EURT", Yekaterinburg (1996). (in Russian).

[8] E.N. Karavaeva and I.V. Molchanova, *The Soviet J. of Ecology*, N10 (1979).

[9] A.S. Bakurov, G.P. Shejn, G.M. Aksenov and S.I. Rovny, Proceedings of the conference on the 50th anniversary of the Mayak accident, 25-26 Sep. 2007 (Chelyabinsk, Russia) pp8-25. (in Russian).

[10] A. Aarkrog, H. Dahlgaard, S.P. Nielsen, A.V. Trapeznikov, I.V. Molchanova, V.N. Pozolotina, E.N. Karavaeva, P.I. Yushkov and G.G. Polikarpov, *Sci. of the Total Environ.*, 201 (1997).

[11] A.V. Zheleznov, N.B. Zheleznova, N.I. Smetanin and V.S. Sukhanovskaya, *Russian J. of Genetics*, 38, N5 (2002).

[12] B.V. Nikipelov, A.S. Nikiforov, O.L. Kedrovsky, M.V. Strakhov and E.K. Drozhko, "Practical rehabilitation of territories contaminated as a result of implementation of Nuclear Material Production Defense Programs". Oak Ridge National Laboratory, USA (1990).

[13] G.V. Gorshkov, "Penetrating radiations from radioactive sources," Leningrad (1967). (in Russian).

[14] F.A. Tikhomirov and R.T. Karaban, "Ecological after-effects of the radioactive contamination at South Ural", Moscow (1993) pp85-95. (in Russian).

[15] E.G. Smirnov, "Ecological after-effects of the radioactive contamination at South Ural", Moscow, pp79-84 (1993). (in Russian).

[16] G.M. Woodwell, *Science*, 156, 774 (1967).

[17] R.B. Platt, "Radioecology". Reinhold: N.Y. (1963).

[18] G.M. Woodwell and L.N. Miller, *Science*, 139, 3551 (1963).

[19] F.A. Tikhomirov, "The ionizing radiation influence on ecological systems", Moscow (1972). (in Russian).

[20] N. Poinsot-Balaguer, R. Castet and E. Tabone *J. Environ. Radioactivity*, 14, (1991).

[21] E.G. Smirnov, "Ecological after-effects of the radioactive contamination at South Ural", Moscow, pp103-108 (1993). (in Russian).

[22] A.V. Lagunov, and A.I. Smagin, *Voprosy radiatsionnoi besopasnosti.* Special issue (2007). (in Russian).

[23] L.V. Cherezhanova, R.M. Aleksakhin and E.G. Smirnov, *Russian J. of Genetics*, 7, N4 (1971).

[24] V.A. Shevchenko, V.I. Abramov and V.L. Pechkurenkov, "Ecological after-effects of the radioactive contamination at South Ural", Moscow (1993). (in Russian).

[25] V.A. Shevchenko, V.A. Kal'chenko, V.I. Abramov, A.V. Rubanovich, V.V. Shevchenko, L.I. Grinikh, *Radiatsionnaia biologiia. Radioecologiia*, 39, N1 (1999). (in Russian).

[26] E.A. Lysenko, V.I. Abramov and V.A. Shevchenko, *Russian J. of Genetics*, 36, N9 (2000).

[27] V.A. Kal'chenko, V.I. Abramov, A.V. Rubanovich, L.N. Kostina, I.S. Fedotov, E.V. Igonina, G.I. Kuznetsova, and V.A. Shevchenko, *Radiatsionnaia biologiia. Radioecologiia*, 42, N6 (2002). (in Russian).

[28] V.N. Pozolotina, I.V. Molchanova, L.N. Mikhailovskaya and E.V. Ul'yanova, *Russian J. of Ecology*, 36, N5 (2005).

[29] V.N. Pozolotina, E.V. Antonova and E.M. Karimullina, *Russian J. of Ecology*, 41, N6 (2010).

In: Radiation Exposure
Editors: D. Balenovic and E. Stimac
ISBN 978-1-62100-869-9

Chapter 5

RADIATION EXPOSURE IN PREGNANCY: APPROPRIATENESS, RISKS, DOSE REDUCTION STRATEGIES

Mahesh K. Shetty[1, 2,*]

[1]Clinical Professor of Radiology, Baylor College of Medicine, Houston, Texas, U. S.

[2]Woman's Hospital of Texas, Houston, Texas, U. S.

ABSTRACT

Introduction: This chapter focuses on the topic of Radiation exposure during Pregnancy. The indications for performing radiological procedures to aid in the clinical management of the pregnant patient and the appropriate triaging of the diagnostic modalities are discussed. A brief description of the salient features of Radiation biology as well as effects of radiation on the fetus and the mother is presented. Finally dose reduction strategies available to minimize risk to the fetus, and the need for and the process of obtaining informed consent and counseling the pregnant patient are discussed.

Clinical Indications and Imaging Algorithms: Common indications for performing radiological procedures during pregnancy include diagnosis of Pulmonary Embolism, evaluation of acute abdominal pain for conditions such as acute appendicitis, urolithiasis, and biliary colic as

---

* Tel: 1 713 7917115, Fax: 1 877 3735490. mshetty@womens-radiology.com.

well as assessment of traumatic injuries in a pregnant patient. Fluoroscopy for guidance of interventional procedures and the very rare need for radiation treatment of cancer in pregnancy are additional sources of medical needed radiation exposure in pregnancy.

Radiation Biology, Fetal Radiation Risks, Maternal Radiation Risks: There are threshold effects resulting from exposure to radiation in utero and these include birth defects, mental retardation, neurobehavioral effects, and growth retardation and embryonic death. Radiation induced carcinogenesis is a non threshold phenomena. An example of radiation risk to the mother includes an increased risk for breast cancer resulting from Chest CT performed for diagnosis of pulmonary embolism.

Radiation Dose Reduction Strategies: There are various strategies that can be put in place to reduce radiation dose resulting from Computed Tomographic examinations. Fluoroscopy performed in conjunction with urgent and emergent interventional procedures is an important source of radiation and although less often utilized compared to CT, it is important to adopt techniques to keep the radiation dose as low as possible during fluoroscopy.

Counseling and Consent: The American College of Radiology recommends that written informed consent be obtained in all instances when a pregnant woman is exposed to medically indicated radiation. Counseling of the patient of the risks and the benefits of the procedure is an important component of the consent process. Likewise referring clinicians also need to be educated on the potential hazards and appropriateness of use of radiation during pregnancy.

Conclusion: Exposure to radiation in pregnancy may be medically indicated for an accurate and timely clinical diagnosis, to guide interventional procedures and very rarely to treat cancers. Medical necessity in each instance must outweigh the potential risk to the fetus. Whenever feasible, exposure to radiation must be deferred, replaced by methods that do not use ionizing radiation or when used, must utilize the lowest possible dose so as to keep the deleterious effects to the fetus and the mother at a minimum.

INTRODUCTION

This chapter focuses on the topic of exposure to medical radiation during Pregnancy. The indications for performing radiological procedures in a pregnant patient and the appropriate triaging of the diagnostic modalities are discussed. Pertinent aspects of radiation biology as well as effects of radiation on the fetus and the mother are presented. Finally dose reduction strategies available to minimize risk to the fetus, and the need for and the process of obtaining informed consent and counseling the pregnant patient are discussed.

Common indications for performing radiological procedures during pregnancy include evaluation of acute abdominal pain for conditions such as acute appendicitis, urolithiasis, and biliary colic, diagnosis of Pulmonary Embolism, as well as assessment of traumatic injuries in a pregnant patient. Fluoroscopy for guidance of interventional procedures and the very rare need for radiation treatment of cancer in pregnancy are additional sources of medically necessitated radiation exposure in pregnancy. There are threshold effects resulting from exposure to radiation in utero and these include birth defects, mental retardation, neurobehavioral effects, growth retardation and embryonic death. Radiation induced carcinogenesis is a non threshold phenomena. An example of radiation risk to the mother includes an increased risk for breast cancer resulting from Chest CT performed for diagnosis of pulmonary embolism.

There are various strategies that can be put in place to reduce radiation dose resulting from radiological procedures particularly Computed Tomographic examinations. Fluoroscopy performed in conjunction with urgent and emergent interventional procedures is an important source of radiation and although less often utilized compared to CT, it is important to adopt techniques to keep the radiation dose as low as possible during fluoroscopy. It is recommended to obtain a written informed consent in all instances when a pregnant woman is exposed to medically indicated radiation. Counseling of the patient of the risks and the benefits of the procedure is an important component of the consent process. Likewise referring clinicians also need to be educated on the potential hazards and appropriateness of use of radiation during pregnancy.

Clinical Indications and Imaging Algorithms

Accurate and timely diagnosis of medical conditions that may complicate pregnancy necessitates utilization of appropriate imaging modalities. It is standard clinical practice to avoid using procedures that involves exposure of the developing fetus to ionizing radiation, using instead modalities such as ultrasound and MRI that do not involve radiation. Use of radiological procedures is generally avoided in the first trimester due to heightened risks to the fetus during organogenesis. Whenever procedures that involve use of radiation are needed, efforts to keep the radiation dose as low as reasonably achievable are employed. Whenever it is feasible, direct exposure of the fetus to radiation should be avoided. Radiological procedures performed in a pregnant patient falls under two categories. Diagnostic studies are performed for a variety of indications primarily in the evaluation of acute diseases or trauma. In most instances, the fetal radiation dose and hence deleterious effects on the fetus as a result of diagnostic procedures is seldom of clinical significance. Therapeutic procedures which involve fluoroscopic guidance for interventional procedures and radiation therapy to treat cancer in a pregnant woman may expose the fetus to clinically significant doses.

There has been a trend of increased usage of Computed Tomographic examinations in pregnancy. A recently published study looked at a data base of 170,000 chests and 180,000 abdominopelvic CT scans. There was an increase in performance of chest CT of 75% /year/1000 deliveries in pregnancy compared to 19%/y increase in all patients. Abdominopelvic CT utilization showed an increase of 22%/y in pregnant patients vs. 13%/y for all patients [1].

Clinical evaluation of a pregnant patient is more challenging than in a non pregnant patient. The most common diagnostic indications for non obstetric imaging during pregnancy include abdominal pain, trauma and pulmonary embolism. The most common indication for performing an abdominopelvic CT in pregnancy is suspected appendicitis, accounting for 58% of cases in one series, the commonest indication for a chest CT is in women with suspected pulmonary embolism, accounting for 85% of chest CT's in the same study[1]. The following describes the medical indications that account for most of the radiological imaging performed in a pregnant patient.

Abdominal pain is often non specific and may not always localize to the expected site of disease secondary to displacement of the abdominal viscera by the gravid uterus. Possible causes for pain can include biliary disease, acute appendicitis, and Ureteral calculi. Sonography must always be the initial modality used in the diagnostic work up and readily diagnoses biliary disease. Appendicitis is the most common cause of a surgical intervention for an abdominal condition in pregnancy with a reported rate of 50 to 70 per 1000 patients [2, 3]. Ultrasound is the preferred imaging modality (Figure 1). When inconclusive or technically limited by patient factors, use of CT or MRI is needed. However MRI may not be available and when available may not be an option for use after hours [4]. The need for a prompt and accurate diagnosis will require use of CT (Figure 2). In general, CT is the preferred modality for the evaluation of acute appendicitis in the second and third trimester whenever ultrasound is inconclusive. In the first trimester MRI is preferred if available when ultrasound is inconclusive in the evaluation of suspected appendicitis.

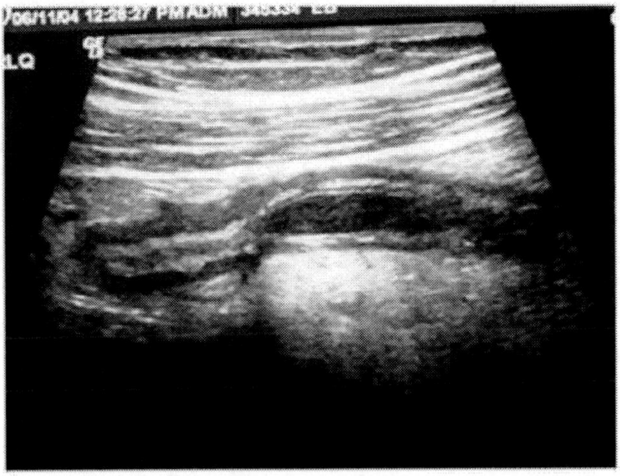

Figure 1. Ultrasound in a pregnant patient with fever and right lower quadrant pain shows a markedly thickened inflammmed appendix.

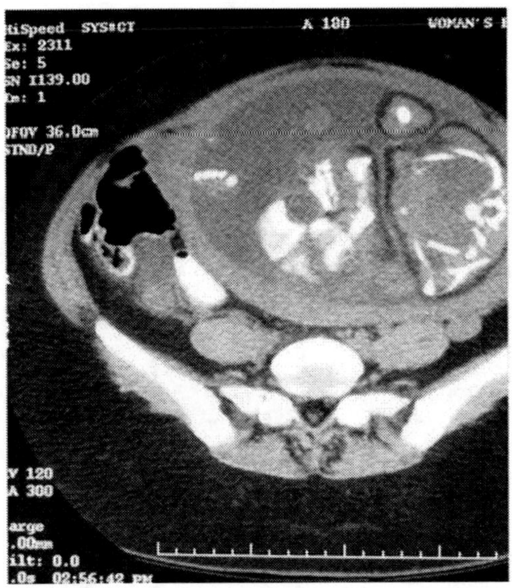

Figure 2. CT scan image in a pregnant patient with suspected appendicitis and negative ultrasound examination demonstrates a thickened inflamed appendix posterior to the cecum. At surgery appendicitis was confirmed.

a)

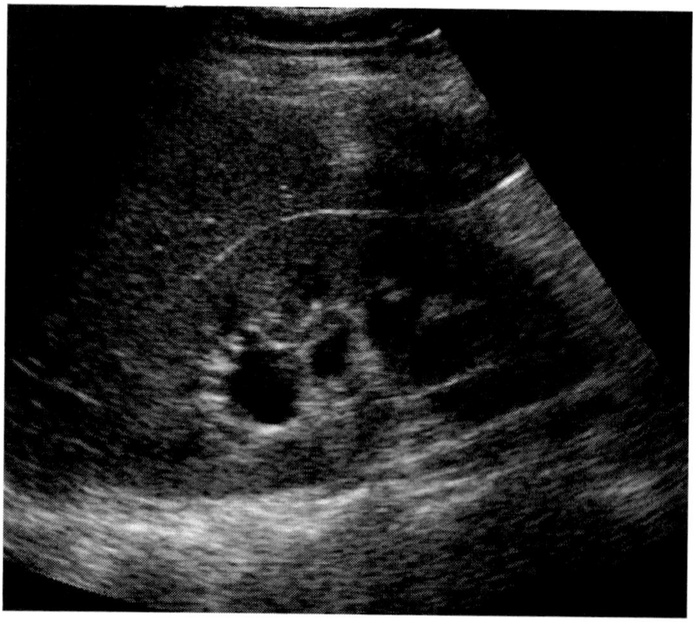

b)

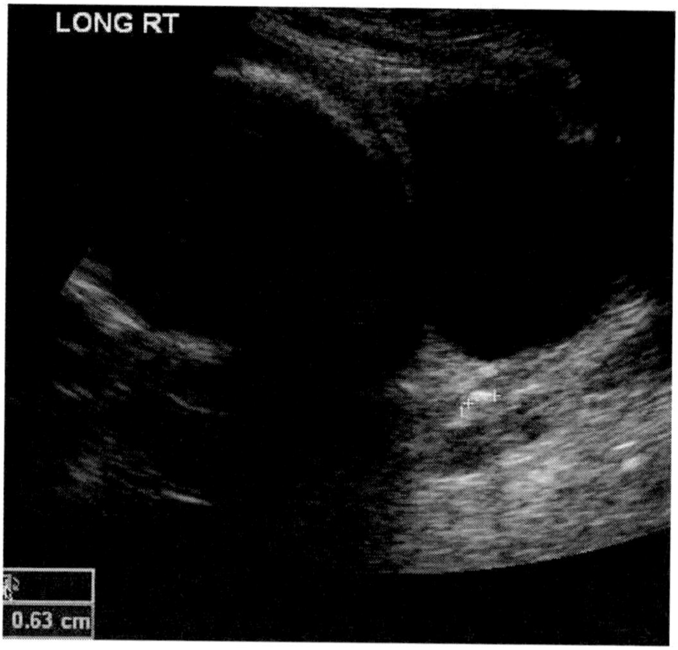

c)

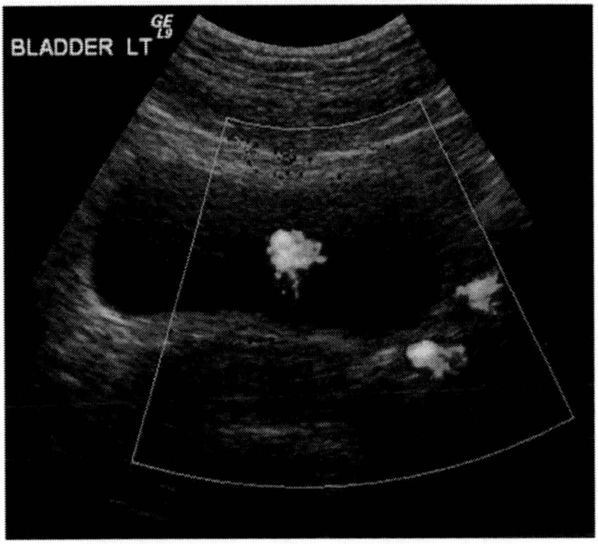

Figure 3. A pregnant woman presenting with acute flank pain in the third trimester. a) Ultrasound image of the right kidney demonstrates hydronephrosis. b) calculus was identified in the distal right ureter [Cursors]. c) Color Doppler Imaging of the Urinary bladder demonstrates a right Ureteral jet and an absent Ureteral jet. A small calculus is seen in the left ureter close to the ureterovesical junction.

The other cause of abdominal plan is urolithiasis; and is the most common painful non obstetric condition in pregnancy. Clinical and sonographic diagnosis has limitations. Pregnancy induced hydronephrosis can mimic obstructive hydronephrosis and thereby causes a dilemma in diagnosing calculus induced obstructive hydronephrosis. Renal Calculi in pregnancy are uncommon, with a prevalence of 0.4 to 5 per 1000 pregnancies; it is more common in the multiparous [5]. Not all calculi may be symptomatic accounting for the wide reported range [5]. In a pregnant patient with ureteral calculus, persistent pain and obstruction and infection are indications for intervention. Conservative treatment consists of rest, intravenous fluids and analgesics. Surgical intervention is usually not needed since 75% of stones pass spontaneously. [5]. Ultrasound is always the initial imaging modality but as mentioned before is complicated by physiologic pregnancy induced hydronephrosis particularly on the right side (Figure 3a). Ureteral calculi are difficult to identify on ultrasound. We routinely use endovaginal scanning in patients with hydronephrosis and acute flank pain to identify distal Ureteral calculi (Figure 3b). Ureteral jets are another way of assessing for ureteral obstruction. This involves performing color Doppler imaging of the urinary bladder (Figure 3c). An absent ureteral jet has a reported sensitivity of 100% and a specificity of 91% in the diagnosis of Ureteral obstruction. However a significant percentage of pregnant women without Ureteral obstruction may also have an absent ureteral jet. [6, 7].

Whenever ultrasound is inconclusive a limited IVP can be performed, but this study is often of limited diagnostic valve. The need to limit radiographic views and the fact that the fetal skeleton can obscure a radio opaque calculus are some of the drawbacks of interpreting an Intravenous pyelogram study in pregnancy. Low dose CT is being increasingly used for this indication. It is an accurate way of identifying Ureteral calculi [8, 9]. Untreated ureteral calculus may be complicated by pyelonephritis and hence carries an increased risk of premature labor necessitating intervention. Intervention in such cases has traditionally included placement of ureteral stents or nephrostomy as a temporary measure with definitive treatment deferred until after delivery.

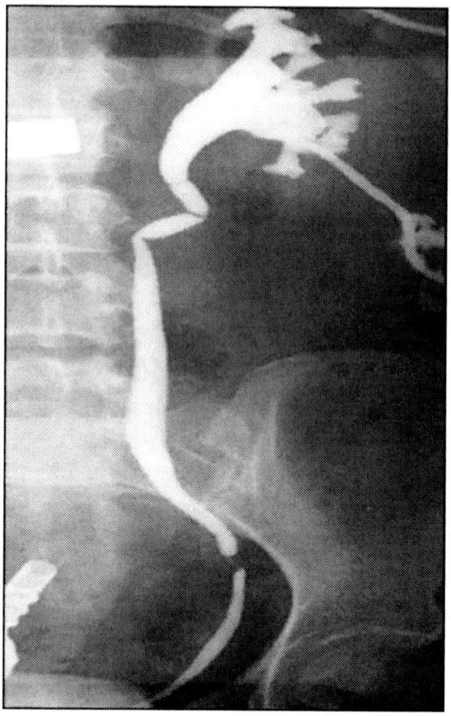

Figure 4. Left Nephrostomogram demonstrates a tight distal Ureteral stricture in a symptomatic pregnant patient, a Ureteral stent was subsequently inserted.

Nephrostomy and or placement of stents is generally performed under fluoroscopy however in a pregnant patient ultrasound or ureteroscopic placement of stents are alternatives to be considered keeping radiation safety in mind. [10, 11] (Figure 4). It has been suggested that definitive treatment of stones can be performed instead of temporary measures during pregnancy. There have been reports of successful ureteroscopic removal in symptomatic pregnant women with an imaging identified calculus obstruction. This method may be a better alternative to nephrostomy or stent placement in those with persistent pain or obstruction [12].

Abdominal Trauma is another important indication for imaging in pregnancy. Reported incidence of accidental injury during pregnancy is 70 per 1000 pregnancies, these are most common in the third trimester. Two thirds of traumatic injuries results from motor vehicular accidents. [13, 14]. The most common injury is placental abruption seen in up to 40% following severe blunt abdominal trauma and in 3% of minor abdominal trauma. Ultrasound is adequate for diagnosis of this condition. [15]. Ultrasound is not only helpful in evaluating the placenta and fetus, it can also detecting maternal large-volume hemoperitoneum and significant solid organ injury. However, bone, viscus, and retroperitoneal injuries are often unnoticed at US (15, 16). CT is the most accurate and cost-efficient diagnostic tool available for evaluation of the hemodynamically stable, seriously injured pregnant patient with blunt abdominal trauma (15,16) It has been proposed that ultrasound be used in trauma patients with a low level of trauma and a low likelihood of injury. In such scenarios additional imaging is performed as needed. These may include CT of the maternal head, chest, abdomen, and pelvis. MR imaging is the modality of choice for neural injuries. No imaging or interventional study should be withheld because of the fetal radiation risk. Angiography may be indicated if active internal bleeding is suspected. [12]

Pulmonary Embolism in pregnancy: Pregnancy attributed Hypercoagulable state and increased venous stasis causes and increased risk of venous thromboembolism reported to be five times greater than in the non pregnant woman [17]. An increased mortality rate from PE is also reported in pregnancy, this combined with increased maternal and fetal morbidity resulting from treatment with anticoagulants means that accurate diagnosis is critical for appropriate management of the pregnant patient.[18,19]. Clinical symptoms and signs are non specific and D-dimer values are not helpful in the pregnant patient. Initial evaluation is by means of compression US of the lower extremities. Finding of a DVT avoids the need for additional testing [20]. The limitations of ultrasound are however its low sensitivity and failure to detect iliac vein DVT that is more commonly seen in the pregnant patient [18]. One study reported that 10% of patients with a high clinical suspicion of PE and negative compression US of the lower extremity had angiographically proven PE, justifying the need for additional testing if clinical suspicion is high. [19]. CT Pulmonary Angiography [CTAP] has a higher sensitivity [81-91%] and specificity [93-97%] than ventilation perfusion scanning in detection of PE in main, lobar and segmental branches of the pulmonary artery. [19] [Figure 5]. This compares to sensitivity with a high probability ventilation-perfusion scan of only 41% [22]. Radiation dose to the fetus from CTPA is similar to or lower than that from a VQ scan. [17, 21, 22]

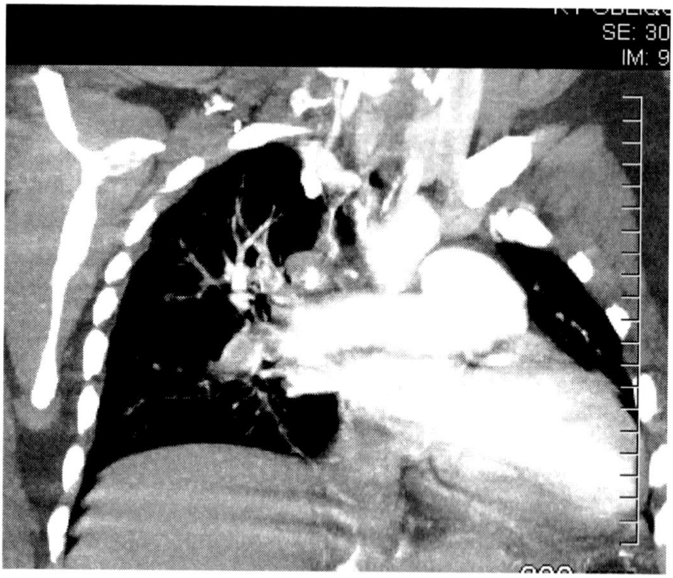

Figure 5. CT Pulmonary angiogram in a pregnant patient with suspected Pulmonary embolism confirms emboli appearing as filling defects in the right ascending and descending segmental branches of the Right pulmonary artery.

Interventional Radiology and Fluoroscopy

During pregnancy fluoroscopy is uncommonly performed and when used is generally used to provide guidance for placement of ureteral stents or for endovascular interventional procedures such as placement of caval filters or placement of balloons in uterine vessels to control bleeding.[23,24]. Placements of ureteral stents may be performed under ultrasound guidance obviating the need of exposure of the fetus to radiation. However all of the other procedures noted above require fluoroscopic guidance. Diagnostic angiography may on occasion be performed in pregnant patients with pelvic trauma and a hematoma to identify active bleeders. Emergent coil embolization of life threatening ruptured renal and splenic arterial aneurysms has performed on pregnant patients has also been reported in the literature. Placement of a stent in an occluded renal artery with malignant hypertension is an additional reported indication. The fetal dose resulting from such a radiological procedure was 0.002 Gy. [25]. whenever fluoroscopy is utilized steps have to undertaken to minimize radiation exposure to the fetus. These steps are outlined by Marx in his article [Table 1] [26].

Table 1. Methods to reduce absorbed radiation dose from fluoroscopy during an interventional procedure

Minimize fluoroscopy time by closely monitoring the time that the fluoroscopy is turned on
Maintain fluoroscopic frame rate at the minimum feasible level
Last image hold is a feature which allows images to persist after the foot is taken of the pedal, and decreases the need for radiographic procedure
When performing Digital subtraction angiography, the DSA frame rate and total number of frames in each sequence should be optimized to the examination being performed. The goal should be to obtain as few images as possible.
In order to minimize absorbed dose, patient is best imaged from a straight postero-anterior projection, this is preferred to a lateral or oblique projection.
The image receptor should be as close to the patient as possible and the x-ray tube should be as far away from the patient as possible
Field collimation should be as tight as possible
Magnification mode usage should be kept at a minimum.
Patient shielding of the region not being imaged is important and the shield should be placed between the patient and the x-ray tube
It may be advisable to have a physicist determine the absorbed fetal dose after completion of an interventional procedure

Source Reference number [26].

Uncontrolled bleeding as in placental implantation abnormalities may require arterial embolization procedures. This is a life threatening situation and saving the pregnant woman's life is paramount. There are other emergent and urgent clinical indications where management may be unchanged irrespective of the pregnancy status. These indications may include biliary sepsis, sepsis secondary to Ureteral calculus obstruction, blunt or penetrating trauma with internal hemorrhage, aortic dissection, pulmonary embolism, venous access. On the other hand elective interventional radiological procedures can be deferred until after pregnancy. However a procedure such as elective closure of a pulmonary AVM may be best done promptly due to increased risk of an untreated AVM resulting from pregnancy. It is important to include both the patient and the obstetrician in the decision making process in all cases. [26]

Radiation Dose Considerations

Radiation dose and radiation risk resulting from a radiological examination is expressed as the absorbed dose that is measured in grays (Gy); the effective dose equivalent, measured in sievert (Sv), the CT dose index (CTDI) measured in mGy and dose length product (DLPI) measured in mGy.cm. The CTDI is the most commonly used dose indicator for CT examinations. It is measured and calculated in a polymethylacrylate phantom, and measured in mGy in a single slice. It's limited by not accounting for patient's size. The value applies to scanning at a pitch of 1. Dividing the dose index by pitch provides volume CTDI. DLP provides an estimate of the radiation dose of an entire examination by taking into account the total scan length or the Z factor. The CTDI and DLP provide an approximate estimate of the radiation dose resulting from CT scan examinations and are available on the console. Although not a precise estimate of the radiation dose, an idea of the variation of the dose when the technical parameters are changed provides an idea of the benefit in reduction of radiation dose after changes in the pitch and or tube current. [27, 28]

The best available parameter to estimate the total radiation risk to the patient is the effective dose that is expressed in Sv. This is obtained by summing up the absorbed doses of individual organs weighted for organ sensitivity. There are various computer programs that can calculate dose for individual organs using the volume CTDI and organ weighting factors from International Commission on Radiological Protection publication 60 [29], it's not possible to determine the fetal radiation dose directly; existing methods are on the basis of phantom measurements and or geometrical phantom simulation methods. These assume early term pregnancy in a single-size patient model with an average non varying maternal anatomy. A more accurate method of estimating fetal radiation dose resulting from an abdominal and pelvic examination for a range of gestational age and patient size has been reported. [30]

These investigators used Monte Carlo simulations with a multidetector CT scanner to estimate fetal dose resulting from abdominal pelvic CT examinations for a range of gestational age and patient sizes. Fetal dose correlated with maternal perimeter and varied more than thought previously reported. When both maternal size and fetal depth are combined correlation improves further. Maternal perimeter was chosen due to the ease of estimation using a tape measure around the estimated center of the fetus. The fetal depth also correlated well with the radiation dose, it was measured using ultrasound or from the CT image as the distance from the skin surface to the fetus. There was no correlation between fetal radiation dose and gestational age at which the examination was performed. However normalized fetal dose was reported to decrease linearly with increasing patient perimeter. Among the 24 patients used as study models, the range of fetal doses varied from 7.3 to 14.3 mGy/100 mA. Although the Monte Carlo stimulations used by the authors relate to a single set of acquisition parameters, their results can be used to estimate the radiation dose for differing technical parameters using the equation detailed in their manuscript [30]. As an overview, it's useful to know that the annual background radiation dose is about 2.4 mSv/year, fetal dose from CT or fluoroscopy when fetus is not in the x-ray beam is typically less than 5 mSv. Fetal dose from abdominal radiography, lumbosacral spine examination, and limited intravenous urogram is typically less than 10 mSv. CT of the abdomen and pelvis, and interventional procedures involve fetal exposures of greater than 10 mSv. Table 2 provides an overview of typical values for radiation doses resulting from commonly performed radiological procedures. Table 3 lists the dose resulting from commonly performed Computed Tomographic examinations.

Table 2. Effective Doses in Computed Tomographic and Radiographic Examinations

Computed Tomographic Examination	Effective Dose (mSv)	Radiographic Examination	Effective Dose (mSv)
Head	2	Skull	0.07
Chest	8	Chest	0.02
Abdomen	10-20	Abdomen	1.0
Pelvis	10-20	Pelvis	0.7
		Ba Swallow	1.5
		Ba Enema	7

ICRP publication 87: Managing Patient Dose in Computed Tomography.

Table 3. Estimated Average Fetal Radiation Doses from a Single Acquisition with a 64-Row Multidetector Volume CT Scanner

	CT Protocol			Imaging Parameters	
Type of CT Examination	Dose (mGy)	Section Thikness (mm)	Noise Index	Tube Current-Time Product (mAs)	Pitch
CT of the chest	0.02	2.5	30	80	1.375
CT pulmonary angiography	0.02	1.25	30	88	0.984
CT of the abdomen	1.3	2.5	36	110	1.375
CT of the kidne, urater, and bladder	11	2.5	36	110	1.375
CT of the pelvis	13	2.5	36	130	1.375
CT of the abdomen and pelvis	13	2.5	36	130	1.375
CT angiography	13	2.5	30	130	1.375

Reproduced with Permission [12].

Biological Effects from Exposure to Radiation

The risk to the fetus from exposure to ionizing radiation is dependent on the stage of development, i.e., stage of pregnancy and the radiation dose. During the first trimester of pregnancy when organogenesis is occurring the risks are the greatest and are the least during the third trimester. Radiological studies that do not include the abdomen and therefore the fetus have clinically insignificant effect on the fetus. Therefore, the most attention needs to be focused on those studies where the developing fetus is in the direct field of view, i.e., abdominal–pelvic CT. Whenever possible physicians should consider utilizing alternate methods that do not use ionizing radiation. The effects of exposure to ionizing radiation have to be considered in the context of certain preexisting background risks that is common to all pregnancies. These include a 3% risk for birth defects, 15% for miscarriage, 4% for prematurity, 4% for growth retardation, and 1% risk of mental retardation or neurological developmental problems. [4]

The damaging effects of Ionizing radiation are due to cell death and deoxyribonucleic acid changes, this forms the basis of all biological effects of radiation on the developing fetus. Cell death leads to morphologic changes that cause growth retardation, congenital malformations, mental retardation, and in the extreme instance still birth. [4] Genetic mutations and carcinogenesis result from Deoxyribonucleic acid changes. Morphologic changes results from the destructive effects of radiation involving many cells and is hence a threshold phenomenon. Carcinogenesis on the other hand may be initiated by injury to a single cell. Radiation effects are therefore of two types, one is deterministic and the other is stochastic. Deterministic effects are dose related; there is a baseline threshold dose above which the severity of the radiation induced changes increases with increasing dose. Such changes result from damage to multiple cells. Deterministic effects do not occur at doses below certain threshold levels. Stochastic effects result from radiation induced damage in single cells, and form the basis for increased risk of carcinogenesis and genetic changes that result from damage to reproductive genes.[4] These changes have no threshold but the likelihood increases as the dose increases. Among the known risks resulting from radiation exposure the thresholds for deterministic effects, such as congenital malformations, growth retardation, pregnancy loss, and neurobehavioral effects are all greater than 200 mGy [31] The threshold for mental retardation is about 180 mGy during 8-15 weeks of gestation and is higher in the later period of pregnancy.[31,30] The risk of increased incidence of cancer be it leukemia or solid cancers in childhood resulting from inutero exposure to radiation is difficult to assess, and is not a threshold phenomenon. The precise relationship between radiation and carcinogenesis is unclear. A causal relationship has been made only on the basis of case control studies; cohort studies have not shown any association, including in atomic explosion survivors irradiated in utero. Leukemogenic risk seems to be more likely than other cancers but the risk is hard to quantify [9] It has been reported that the lifelong fatal cancer risk for every 10, 000 infants born each year is 18%, although all 10,000 received a dose of 10 mSv, there would be 1 excess case of cancer in the population of 10, 000. [30, 31] The cancer risk from diagnostic radiological procedures is insignificant when compared with the spontaneous risk of cancer. [32,33]. It has been shown that there is an increased risk of cancer following exposure of the adult population to high doses of radiation such as those who survived the atomic bomb, however the magnitude of the risk of cancers from embryonic exposures to diagnostic radiological procedures is very controversial. [34] In fact recent reports suggest that the risk may be lower for the irradiated embryo than the irradiated child and that there may be no increased risk of carcinogenesis from diagnostic radiological procedures [35]. Lilienfeld published a review of epidemiological studies on leukemogenesis and has concluded that results are consistent in showing an excess frequency of leukemia in children of mothers who were exposed to radiation in pregnancy[36]. However in a study of 1300 people some of whom were exposed to the atomic bomb in utero, there was no increased evidence of malignancy in 24 years of follow up [37], there was however an increased mortality in the first year of life and after 10 years. Another study of 17000 children of parents that were exposed to the atomic bomb before conception found no increase in the incidence of leukemia [38]. Brent opines that although there are some investigators who believe that in utero exposure to small amounts of radiation increases the risk of leukemia and other malignancies, there are others who are questioning whether the embryo is indeed any more sensitive to the carcinogenic effect of irradiation than a child or an adult.[35] In a study of the occurrence of cancers in adults who were irradiated in utero in Hiroshima and Nagasaki it was shown that there was no statistical increase in the oncogenic risks when in utero exposures were less than 20 rads [0.2 Sv]. The in utero population was much less sensitive to the oncogenic effects of radiation than the children who were exposed to the atomic bomb. [39]. Table 4 summarizes biological effects of radiation at different stages of gestation.

Table 4. Summary of Suspected In-Utero Induced Deterministic Radiation Effects *[40]

Menstrual Age	Conception Age	< 50 mGy	50-100 mGy	100 mGy
0 -2 wk	Prior to conception	None	None	None
3rd-4th week	1st-2nd week	None	Probably none	Possible spontaneous abortion
5th -10th week	3rd-8th week	None	Potential effects are scientifically uncertain and probably too subtle to be clinically detectable	
11-17th week	9-15th week	None	Potential effects are scientifically uncertain and probably too subtle to be clinically detectable	
18-27th week	16th-25th week	None	None	IQ deficits are not detectable at diagnostic dosed
>27 weeks	>25 weeks	None	None	None applicable to diagnostic medicine

* ACR Practice guideline for imaging pregnant or potentially pregnant adolescents and women with ionizing radiation. ACR practice guideline for continuing medical education. Practice Guidelines and Medical Standards. Reston, Va: American College of Radiology 23-37, 2008.

Dose Reduction Strategies

The ACR practice guidelines for imaging pregnant or potentially pregnant adolescents and women with ionizing radiation states that ' to maintain a high standard of safety, particularly when imaging potentially pregnant patients, imaging radiation must be applied at levels as low as reasonably achievable (ALARA), while the degree of medical benefit must counterbalance the well-managed levels of risk." [40]

Abdominal CT

The need for radiation dose reduction is of particular interest when the developing fetus is directly exposed to ionizing radiation as in an abdominal pelvic CT. Scanning only the area of clinical interest and reducing the number of phases to the absolute minimum often 1 phase only are ways of minimizing the radiation dose. The availability of automatic tube current modulation wherein the tube current is decided on the basis of patient size and density on the scanogram additionally provides the radiologist the option of choosing parameters that can be tailored to the clinical question pertaining to the patient being scanned. A higher noise index resulting from a low dose CT often provides adequate diagnostic in situation where sharpness of lesions is not critical, such as required in the evaluation and or characterization of renal or hepatic masses. For some of the common indications for scanning a pregnant abdomen namely, appendicitis, abdominal abscess, and calculus disease, an image using low dose protocol is feasible particularly because a higher noise Index would still provide the clinical information desired in evaluation for appendicitis, abscess, or calculus disease. [4]

A low-dose CT protocol that involves reducing the scan range, decreasing the tube current, and increasing the pitch in comparison with those of the standard protocol is generally suited in the pregnant patient. In some cases, it may be possible to reduce the kilovoltage without compromising image quality.[12]. Although fetal dose resulting from commonly performed medical imaging procedures is generally clinically insignificant, one needs to use caution when multiple exams are performed. Radiation dose to the fetus is most concerning after multiple consecutive studies have been performed since the accumulation of radiation dose may then reach the threshold dose. [12] For imaging of Ureteral calculi fetal radiation doses of 4 to 7.2 mGy and 8.5 to11.7 mGy at 0 and 3 months of gestation, respectively has been reported.

The low dose CT protocol in this study uses a relatively low tube current (160 mA, 140kVp) and a 16-row multidetector CT scanner [41]. Dose reduction strategies in CT are outlined in Table 5.

Table 5. Dose Reduction Techniques for CT of Pregnant Patients

One size does not fit all: do not use standard protocols
Decrease kilovoltage for small patients
Decrease milliamperage and use automatic tube current modulation
Increase pitch to >1
Obtain a single scout view and avoid directly imaging the fetus for planning purposes
Limit the field of view
Avoid imaging in multiple phases
Use more recently available novel reconstruction algorithms to reduce noise in images, thus allowing reduction of milliamperage or increase in noise level requirements during scanning
Lead shielding of the mother; most pronounced effect with circumferential shielding
Internal barium shielding with use of oral 30% barium sulfate solution
Local quality assurance program to monitor CT protocols and the resulting dose

Reproduced with Permission [12].

COUNSELING OF PREGNANT WOMEN CONCERNING THE RISK OF RADIATION EXPOSURE IN PREGNANCY

A clear understanding of the biological basis of damage from exposure to radiation, doses resulting from procedures performed on a pregnant patient is a prerequisite for all physicians involved in the decision making process of subjecting a pregnant patient to a radiological procedure. Appropriate counseling requires a thorough understanding of the potent risks to a developing fetus that has been exposed to diagnostic radiation.

Clinicians may not be well aware of the facts relating to the use of diagnostic radiological studies in pregnancy. This may in turn lead to patients having a limited understanding of radiation effects on the fetus causes unnecessary anxiety in pregnant patients exposed to diagnostic radiation. One of the consequent dangers of this lack of knowledge is unnecessary pregnancy termination. In a study that examined physician perceptions of teratogenic risks associated with undergoing plain radiography and CT during early pregnancy, 6 of the 208 family practice physicians stated that they would recommend pregnancy termination after first trimester CT and 1 following radiography in the first trimester. In the same study, 25 out of 208 physicians were not sure of the need for pregnancy termination after radiography; and 39 of the 208 family practice physicians were not sure about a need for pregnancy termination following a CT scan examination. The same study reported that 5 of the 65 obstetricians included in the study would have recommended pregnancy termination after first trimester CT scan examination. [42]

The pregnancy website of the health Physics website provides a recommended source of scientifically sound information pertaining to radiation effects in pregnancy. In 2007 this site had nearly 1.3 million visits. About 8% of the website contacts who had consulted a professional were provided inaccurate information that could have resulted in an unnecessary interruption of a wanted pregnancy [35]. Appropriate counseling should address the small potential radiation risk as well as the substantial and immediate risk of harm that could result if the patient refused to undergo the indicated examination [4]. Nearly all of the diagnostic procedures when performed within the accepted medical standards do not reach the threshold at which birth defects, mental retardation and other neurobehavioral effects, or growth retardation and embryonic deaths occur.

During the first four weeks of pregnancy which represent the preimplantation and pre organogenesis stages of embryonic development, the embryonic cells are least likely to be malformed because these cells are omnipotential and can replace adjacent cells that are damaged. This period is hence referred to as the all or none phenomenon meaning that there may be embryonic death or survival and least likely to be malformed [35].The risk of carcinogenesis is controversial. However when counseling the patient on the small reported risk of carcinogenesis that can occur when exposed to medical radiation, it is worthwhile pointing out that data from Hiroshima and Nagasaki has not recorded a single case of childhood leukemia in the population that was exposed to radiation from the atomic explosion [35]

Pregnancy Imaging Policy

It is advisable that all radiology departments have a policy to guide imaging in pregnancy. Such a policy would aim to create a clear understanding between radiologists, referring clinicians, and pregnant patients undergoing diagnostic radiological examinations. The pregnancy policy would outline the background information needed regarding known effects of diagnostic radiation exposure, as well as expected doses from commonly performed procedures. The policy would put in place steps to identify a pregnant status of every woman of child bearing age who presents for radiological investigations. The counseling of a pregnant patient should be tailored, depending on the area imaged and anticipated exposure dose. Such a policy would also focus on technical aspects of reducing the radiation dose to the patient [43]. The need for obtaining written consent for examination where the absorbed dose is expected to be under 100 mSv has been questioned, given the fact that deterministic effects are nonexistent and the cancer risk in later years is less [34]. A survey of academic centers performing imaging studies on pregnant patients found that 68% of centers obtained informed consent before performing a CT scan examination [44].The American College of Radiology practice guideline for imaging pregnant patient women with ionizing radiation recommends obtaining consent from a patient known to be pregnant and considers it an essential component of providing comprehensive medical care [40]. Furthermore, the guidelines state that it is advisable to document written or verbal consent in the patient's medical record. A written consent should be part of the patient medical record, and in those instances where only a verbal consent has been obtained it should be documented within the Radiology Information System [40]. Given the uncertainty that surrounds the true risk of exposure to radiation and the often conflicting data available in the literature it's prudent to have a consent form that is simple and easy to comprehend. The consent form must be worded in a way that is easy to understand and that conveys to the patient that the risks to the fetus from typical radiation doses that result from abdominal pelvic CT is small, and the benefits resulting from the study far outweighs the minimal risk of cancer resulting from radiation.

Radiation Therapy in Pregnancy

Cancer occurs rarely in pregnancy, the reported incidence of about 1 in 1000. Most frequently diagnosed cancers in pregnancy are breast cancer, cervical cancer, Hodgkin's disease and malignant melanoma [46, 47]. It is general clinical practice to defer radiotherapy in women during pregnancy. However as mentioned earlier stochastic effects have a threshold of 0.1 to 0.2 Gy. If the tumors are away from the fetus and if steps are taken to shield the fetus against leakage radiation and scatter of the teletherapy machine, this threshold is not reached. Cancers in this group would include breast cancer, brain tumors, head and neck cancers, and supradiaphragmatic Hodgkin's disease [47]. However as for diagnostic radiation exposure, the theoretical small risk of carcinogenesis is present as stated previously. Fractionation of the fetal dose, since overall treatment time is long also means that the deleterious effects are less than anticipated. There is no justification to terminate pregnancy when the fetal dose is less than 0.2 Gy [200 rads]. In those instances where a woman is treated with radiation without shielding, which can occur when the pregnant status is unknown at the time of therapy, and the resultant fetal dose, exceeds 500 rads, fetal damage may be substantial. The type and extent of this damage will depend on the stage of pregnancy. Such a dose in the first trimester there may be significant growth retardation and CNS damage. Parents will have to be counseled based on risk assessment by a qualified health physicist allowing them to make decisions on pregnancy termination. [48].Pretreatment dose measurements by a qualified medical physicist are necessary for reliable prediction of side-effects and, hence, adequate provision of information to parents.

Conclusion

Exposure to radiation in pregnancy may be medically indicated for an accurate and timely clinical diagnosis, to guide interventional procedures and very rarely to treat cancers. Medical necessity in each instance must outweigh the potential risk to the fetus. Whenever feasible, exposure to radiation must be deferred, replaced by methods that do not use ionizing radiation or when used, must utilize the lowest possible dose so as to keep the deleterious effects to the fetus and the mother at a minimum. Physicians performing or supervising medical procedures that expose the fetus to radiation need to have a very thorough understanding of the appropriate indications, appropriate methods, steps to keep radiation at a minimum and a knowledge of the potential harms to the fetus so as to allow them to competently counsel pregnant women and inform them of the risks and benefits of the examination that they would be undergoing. A written consent properly documented should be permanent part of the patient medical record. Physicians who request radiological procedures for optimal and timely management of medical and surgical conditions in pregnancy must be aware of the risks and benefits of subjecting their patients to these radiological procedures.

BIBLIOGRAPHY

[1] Goldberg-Stein S, Liu B, Hahn PF, Lee S. Body CT During Pregnancy: Utilization Trends, Examination Indications, and Fetal Radiation Doses. *AJR* 2011; 196:146–151.

[2] Tamir IL, Bongard FS, Klein SR. Acute appendicitis in the pregnant patient. *Am. J. Surg*. 1990;160(6): 571–575; discussion 575–576.

[3] Vu L, Ambrose D, Vos P, Tiwari P, Rosengarten M, Wiseman S. Evaluation of MRI for the diagnosis of appendicitis during pregnancy when ultrasound is inconclusive. *J. Surg. Res*. 2009; 156(1):145–149.

[4] Shetty MK. Abdominal Computed Tomography During Pregnancy: *A Review of Indications and Fetal Radiation Exposure Issues.* Semin. Ultrasound CT MRI 31:3-7.

[5] Evans HJ, Wollin TA. The management of urinary calculi in pregnancy. *Curr. Opin. Urol*. 2001;11(4): 379–384.

[6] Deyoe LA,C ronan JJ, Breslaw BH, Ridlen MS. New Techniques of ultrasound and color Doppler in the prospective evaluation of acute renal disease. Do they replace the Intravenous urogram? *Abdom. Imaging* 1995:20:58-63.

[7] Wachsberg RH. Unilateral absence of Ureteral Jets in the third trimester of pregnancy: pitfall in color Doppler US diagnosis of urinary obstruction . *Radiology*. 1998:209:279-281.

[8] Katz DS, Venkatramanan N, Napel S, Sommer FG. Can low dose unenhanced multidetector CT be used for routine evaluation of suspected renal colic? *AJR Am. J. Roentgenol*. 2003:108:313-315.

[9] Tack D, Sourtzis D, Delpierr I, De Maertelaer V. Gevenois PA. Low-dose unenhanced multidetector CT of patients with suspected renal colic. *AJR Am. J. Roentgenol.*

[10] Semins MJ, Trock BJ, Matlaga BR. The Safety of Ureteroscopy during Pregnancy: A systemic Review and Meta-Analysis. Vol. 181, 139-143, January 2009. *The Journal of Urology.*

[11] Semins MJ, Matlaga BR. Management of Stone disease in Pregnancy. *Current Opinion in Urology* 2010, 20:174–177.

[12] Weiseler KM et al. Imaging in Pregnant Patients: Examination Appropriateness. *RadioGraphics* 2010; 30:1215-1233.

[13] Mattox KL. Goetzl I, Haans L et al. *Crit. Care Med.* 2005: 33(10 SUPPL): S 385-S389).

[14] Goldman SM, Wagner LK, Radiologic ABC's of maternal and fetal survival after trauma: when minutes may count. *RadioGraphics* 1999:19(5): 1349-1357.

[15] Lowdermilk C, Gavant ML, Qaisi W et al. Screening Helical CT for evaluation of blunt traumatic injury in the pregnant patient. *Radiographics* 1999; 19 (spec issue):S243-S255;discussion S256-S258.

[16] Rhea JT, Garza DH, Novelline RA. Controversies in emergency Radiology.CT versus Ultrasound in the evaluation of blunt abdominal trauma. *Emerg. Radiol.* 2004;10(6):289-295.

[17] Winter-Muram HT, Boone JM, Brown HL et al. Pulmonary Embolism in pregnant patients: fetal radiation dose with helical CT. *Radiology* 2002;224:487-492.

[18] Nijkeuter M, Ginsberg JS, Huisman MV. Diagnosis of deep vein thrombosis and pulmonary embolism in pregnancy: a systematic review. *J. Thromb. Haemost.* 2006:4:496-500.

[19] Scarsbrrok AF, Evans AL, Owen AR, Gleeson FV. Diagnosis of suspected thromboembolic disease in pregnancy. *Clin. Radiol.* 2006: 61:1-12.

[20] American College of Radiology. *ACR Appropriateness Criteria for suspected lower extremity deep vein thrombosis.* Available at: http://www.ACR.ORG/S_ACR/BIN.ASP?CID_1208andDID11746andDOC.

[21] Patel SJ, Reede DL, Katz DS, Subramaniam R, Amorosa JK.Imaging the Pregnant Patient for Nonobstetric conditions:Algorithms and Radiation Dose Considerations. *RadioGraphics* 2007; 27:1705–1722.

[22] Nijkeuter M, Geleijns J, De Roos A, Meinders AE, Huisman MV. Diagnosing pulmonary embolism in pregnancy: rationalizing fetal radiation exposure in radiological procedures. *J. Thromb. Haemost.* 2004: 1857-1858.

[23] Tan CH, Tay KH, Sheah K, Kwek K, Wong K, Tan HK, et al. Perioperative endovascular internal iliac artery occlusion balloon placement in management of Placenta accreta, *Am. J. Roentgenolog.* 2007;189:22.1158-63.

[24] Chunilal SD, Bates SM, Venous thromboembolism in pregnancy: Diagnosis, management and prevention. *Thromb. Haemost.* 2009;10:21. 428-438.

[25] Le TT, Haskai ZJ, Holland GA, Townsend R. Endovascular stent placement and magnetic resonance angiography for management of

hypertension and renal artery occlusion during pregnancy. *Obstet. Gynecol.* 1995: 85:822-5.

[26] Victoria Marx M.Interventional Radiology: Management of the Pregnant Patient. *Tech. Vasc. Interventional Rad.* 13:154-157.

[27] Lee CH, Goo JN, Lee HJ, et al: Radiation dose modulation techniques in the multidetector CT era: From basics to practice. *RadioGraphics* 28:1451-1459, 2008.

[28] Frush DP, Applegate K: Computed tomography and radiation: Understanding the issues. *J. Am. Coll. Radiol.* 1:113-119, 2004.

[29] 1990 Recommendations of the International Commission on Radiological protection. *Ann. ICRP* 21:1-201, 1991.

[30] Angel E, Wellnitz C, Goodsit MM, et al: Radiation dose to the fetus for pregnant patients undergoing multidetector CT imaging: Monte Carlo simulations estimating fetal dose for a range of gestational age and patient size. *Radiology* 249:220-227, 2008.

[31] Brent RL: Utilization of developmental basic science principles in the evaluation of reproductive risks from pre- and post conception environmental radiation exposures. *Teratology* 59:182-204, 1999.

[32] Schull WJ, Otake M: Cognitive function and prenatal exposure to ionizing radiation. *Teratology* 59:222-226, 1999.

[33] Boice JD, Miller RW: Childhood and adult cancer after intrauterine exposure to ionizing radiation. *Teratology* 59:227-233, 1999.

[34] Brent RL, Mettler FA: Pregnancy policy. *AJR Am. J. Roentgenol.* 182: 819-822, 2004.

[35] Brent RL. Saving lives and changing family histories: appropriate counseling of pregnant women and men and women of reproductive age, concerning the risk of diagnostic radiation exposures during and before pregnancy. *American Journal of Obstetrics and Gynecology*. January 2009.4-24.

[36] Lilienfeld AM. Epidemiological Studies of the Leukemogenic effects of radiation. *Yale J. Biol. Med.* 1966; 39:143-64.

[37] Hoshino T, Itoga T, Kato H. *Leukemia in the offspring of parents exposed to the atomic bomb at Hiroshima and Nagasaki*, Presented at the annual meeting of the Japanese Association of Hematology, Nagasaki, Japan, March 28-30, 1965.

[38] Kato H. Mortality in children exposed to the A-Bombs while in utero. *Am. J. Epidemiol.* 1971; 93:435-42.

[39] Preston DL, Cullings H, Suyama A, et al. Solid Cancer Incidence in atomic bomb survivors exposed in utero or as young children. *J. Natl. Cancer Inst.* 2008; 100:428-36.
[40] American College of Radiology. *ACR practice guideline for imaging pregnant or potentially pregnant adolescents and women with ionizing radiation*. Reston, Va: American College of Radiology, 2008.
[41] Hurwitz LM, Yoshizumi T, Reiman RE, et al. Radiation dose to the fetus from body MDCT during early gestation. *AJR Am. J. Roentgenol.* 2006; 186:871–876.
[42] Rathnapalan S, Bona N, Chandra K, et al: Physicians perception of teratogenic risk associated with radiography and CT during early pregnancy. *AJR* 182:1107-1109, 2004.
[43] El-Khoury GY, Madsen MT, Blake ME, et al: A new pregnancy policy for a new era. *AJR Am. J. Roentgenol.* 181:335-340, 2003.
[44] Jaffe TA, Miller CM, Merkel EM: Practice patterns in imaging of the pregnant patient with abdominal pain: A survey of academic centers. *AJR Am. J. Roentgenol.* 189:1128-1134, 2007.
[45] Donegan WL. Cancer and pregnancy. *CA Cancer J. Clin.* 1983; 33: 194–214.
[46] Pavlidis NA. Coexistence of pregnancy and malignancy. *Oncologist* 2002; 7: 279–87.
[47] Kal, HB. Struikmans, H. Radiotherapy during pregnancy: fact and fiction. *Lancet Oncol.* 2005; 6: 328–33.
[48] International Commission on Radiological Protection. Pregnancy and medical radiation. *Ann. ICRP* 2000; 30: 1–43.

In: Radiation Exposure
Editors: D. Balenovic and E. Stimac
ISBN 978-1-62100-869-9

Chapter 6

Pharmaceuticals Photochemical Reactions

Cecilia M. Peralta[a] and Liliana P. Fernández[a,b]
[a]Instituto de Química de San Luis, Argentina
[b]Área de Química Analítica, Bioquímica y Farmacia, Universidad Nacional de San Luis, San Luis, Argentina

"In memoriam" of Dr. Adriana Masi,
prominent researcher,
dear colleague and friend,
who passed away prematurely,
as a consequence of public insecurity,
killed by a shot in the head at the door of her house.

Abstract

Photolytic degradation can be a major limiting factor in the stability of pharmaceuticals. A drug may be chemically affected by the radiation of a particular wavelength only if it absorbs radiation at that wavelength and if the power exceeds a threshold. Ultraviolet radiation, which has a high energetic level, is the cause of many degradation reactions. If the molecule absorbs radiation, it produces excited states of itself and may react photochemically with other species. When absorbing molecules are not directly involved in the reaction, but they transfer their energy to other reactive molecules, the absorbing substance is called photosensitizer. Due to the fact that many experimental variables may be involved in photochemical reaction, the kinetic may be very complex. The intensity and wavelength of light and the size, shape, composition and color of the container can affect the reaction rate. Photodegradation of some pharmaceuticals follows zero order kinetics; moreover, others are degraded following a kinetic of first order.

This chapter includes:

a) The use of UV radiation for degradation of pharmaceuticals present in industrial waste.
b) The use of UV radiation to catalyze the reaction of omeprazole degradation, and their determination based on the modification in the fluorescent properties.

INTRODUCTION

Taking into account that photochemical degradation can be a major limiting factor in the stability of pharmaceuticals, increasing attention to the problem of the photolability of drugs has recently been given [1-3]. Ultraviolet radiation has a high-energy level and it can be the cause of many degradation reactions [4]. According to ICH [5], the intrinsic photostability characteristics of new drug substances and products should be evaluated to demonstrate that light exposure does not result in unacceptable changes.

Although it is generally possible to protect commercial drug preparation by appropriate packing, occasional exposure before or after commercialization may result in some photoreaction and as a consequence in a significant activity loss or even in the formation of toxic products, the knowledge of the photochemistry of drug substances often helps in predicting or rationalizing phototoxic effects. Furthermore, it is very important that reports on single drugs or groups of drugs are accompanied by mechanistic investigations, where a rationalization of the observed reactivity and of its dependence on changes in the structure and in the medium is offered.

In addition, some examples of important contributions in this way are presented. Fasani et al. [6] performed a mechanistic investigation on the photodegradation of amlodipine, the corresponding besylate and a simple analogue lacking the β-aminoethoxy group using water and organic solvents. Irradiation produced aromatization to the corresponding pyridines.

Photochemical behavior of trimethoprim, a trimethoxybenzylpyrimidine used as bactericide was studied [7]. Direct irradiation caused a slow reaction:

aromatic amines mostly showed little photochemical reactivity and their excited states decay mainly by emission or by internal conversion or inter system-crossing. The molecule resulted quite sensitive to benzylic hydrogen abstraction by n* ketones yielding dimers in the absence of oxygen or oxidized products in its presence.

García et al. [8] investigated the photodegradation of rabeprazole, in order to determine its kinetics and to elucidate the structures of the main degradation products by HPLC. Rabeprazole sodium is a proton pump inhibitor, used in acid-related disorders, like peptic ulcers and gastroesophageal reflux. When the drug was in methanol solution, one main degradation product (benzimidazolone) was formed, following zero-order kinetics. Another isolated products were identified as benzimidazole and [4-(3-methoxy-propoxy)-3-methyl-pyridin-2-yl]methanol. When powdered commercial tablets were exposed to UVC irradiation, they showed the same degradation products along with other unidentified, which appeared as traces; the degradation rate was slower than in solution. The intact tablets were stable after 50 days of exposition to the same light source.

Potential phototoxic of chemicals could be partly predicted by the determination of reactive oxygen species (ROS) from photo-irradiated compounds. In order to establish provisional classification criteria for risk assessment of drug-induced phototoxicity, Onoue et al. performed ROS assay strategy to 39 marketed drugs and 210 drug candidates [9]. Most phototoxic drugs tended to generate singlet oxygen and superoxide. Of all drug candidates tested, 35.2% compounds were identified as phototoxic or likely phototoxic, and all ROS data for these phototoxic compounds were found to be over the threshold value. Furthermore, 46.3% of non-phototoxic drug candidates were found to be in the sub-threshold region. These results verify that ROS assay can be used for screening purposes in the drug discovery stage.

Five main photoproducts belonging of hydrocortisone 21-acetate (HCA) in methanol solution have been isolated and characterized [10]. The first is a stereoisomer of the andro-derivative, the others arise from the opening of the five-membered ring. Researchers also proved that HCA is photounstable in the solid state and in a commercial formulation for topical use, thus confirming the requirements of the Pharmacopeias for light protection of this drug.

PHOTOCHEMICAL FLUORIMETRY

Molecular fluorescence spectrometry represents a very powerful quantification method for organic compounds because high sensitivity and adequate selectivity can be achieved. Nevertheless, its helpfulness is impeded in numerous cases: when intense ultraviolet (UV) radiation is used for exciting, the analyte can suffer photolysis process, competing with fluorescence measurement, representing a serious reproducibility problem. The fluorescence intensity of the analyte changes during the time of measurement [11-13].

In order to reduce these analytical difficulties, new approaches have been proposed using photochemical reactions for improving the selectivity, sensitivity, and reproducibility in fluorimetric detection. Methods are generally called photochemical fluorescence analysis (or photochemical fluorimetry). It consists of the initiation of a photoreaction of a photochemically unstable analyte by means of UV radiation and the formation of photoproducts.

If the photoproduct is less fluorescent than the analyte, the decay of the fluorescence signal is measured [14,15]. In the majority of compounds, the photochemical reaction yields a photoproduct with an enhanced molecular absorption coefficient and a higher fluorescence quantum yield respect to those of the analyte [16-18]. Consequently, the kinetics can be evaluated, resulting in an improvement of the sensitivity of the fluorimetric detection [16-18].

Specific photochemical reagents, such as photosensitizers, have been used to absorb UV light and initiate a photoreaction. It has helped the development of the method by enabling investigators fluorimetrically to detect compounds with only very weak chomophores [19-21] or without any native fluorescence [22].

Photochemical induction of the fluorescent signal in weakly or non-fluorescent analytes has led to the appearance of fluorescence photochemical or photochemically induced fluorescence at room temperature RTPF [23], used for the analysis of substances, both in solution as well as on solid surfaces. Also, it has been combined with dynamic analytical techniques such as high-performance liquid chromatography (HPLC) [18], thin-layer chromatography (TLC) [24-26], and flow injection analysis (FIA) [27]. In the case of HPLC, photochemical fluorescence has been used as a post-column detection system instead of post-column chemical derivatization [18].

Instrumentation

At present, there is practically no commercial instrument available for photochemical fluorescence analysis. In most cases, laboratory-constructed equipment as well as modifications to commercial instruments have been designed by researchers for measuring photochemical fluorescence in various media, and a variety of new photochemical reactors have been constructed.

Photoreactions in Fluorescence Analysis

Many types of photoreactions might be used for photochemical derivatization and enhancement of fluorescence detection. However, the photoreactions must fulfill several important requirements:

- Reagents (analyte molecules) should strongly absorb radiations in the UV-visible region to initiate the photochemical reaction at wavelengths not significantly absorbed by photoproducts.
- Photoproducts should present increased rigidity and/or aromaticity, resulting in an enhancement absorption coefficient and a higher fluorescence quantum yield with respect to analyte.
- Photoproducts should be chemically and thermally stable on the time scale needed for analytical measurements.
- Photoconversion process should be very efficient.

Reactions that can occur from UV radiation are based on induced fluorescence photocyclization, photoisomerization, photolysis, photoreduction and photooxidation reactions. Much of the first lead to the formation of phenanthrene, anthracene or derivatives [28-32], while those of photoisomerization [33] leading to naphthalene derivatives.

Applications of Photochemical-Induced Fluorescence in Pharmaceutical Analysis

Photochemical fluorimetry has been extensively applied to the determination of pharmaceuticals in the pure state and in preparations.

In relation to applications directed to therapeutic drug, there is information found in literature regarding various active ingredients, such as sulphenamides, antimalarials, analgesics, phenothiazines, among others. The following Table 1 brings together the latest methods.

The high precision and the increased sensitivity of photochemistry-fluorescence measurements make it possible to distinguish between small differences in concentration of pharmaceutical at the ng mL^{-1} levels.

Table 1. RTPF applications to analysis of active ingredients in drugs

Compound	Comments	References
AAS, chloroquine and theophyline	Filter paper	34
TRF. 5-hydroxyTRF. 3-hydroxytyramine, quinine, tryptamine, vitamin K. phenothiazines, phenylbutazone, lorazepam	Different solvents and pHs	14, 16, 35, 36, 37
Diethylstilbestrol, estrogen, digoxin, hydrocortisone, tamoxifen, vitamins, antihistamines, clobazam.	HPLC applied to biological samples	36, 20, 21, 32, 38, 39, 40, 41
Chlorpromazine, promethazine, perphenazine. Several phenothiazines.	FIA, Stopped flow applied to drugs FIA. Inclusion complexes	42 43, 44 45,46
Catecholamine isomers	HPLC in urine	47
Ambroxol, AAS,1,4-benzodiazepine mianserin	HPLC in biological samples and plasma	48,49

AAS: Acetylsalicylic acid; TRF: Tryptophan.

A) UV Radiation for Degradation of Pharmaceuticals Present in Industrial Waste

Water contamination by pharmaceuticals is widely studied and documented [50-53]. The presence of trace pharmaceutical and other xenobiotic compounds in finished drinking water is another public health concern, since little is known about potential chronic health effects associated with long-term ingestion of mixtures of these compounds through drinking water. Some of the adverse effects caused by pharmaceutical pollution include aquatic toxicity, resistant development in pathogenic bacteria, genotoxicity, and endocrine disruption.

Pharmaceuticals are often not completely removed in sewage treatment plants. In order to achieve the quality standards required for wastewater discharge, the effluents contaminated by pharmaceutical derivatives must be treated by appropriate physicochemical processes. Recently, a growing interest has been observed in the area of UV-activated processes due mainly to low cost.

Yuan et al. [54] studied the degradation of four pharmaceutical compounds (ibuprofen, diphenhydramine, phenazone and phenytoin) using ultraviolet (UV) photolysis and UV/H_2O_2 process with a low-pressure (LP) UV lamp. The second-order rate constants of the reaction between the pharmaceuticals and -OH are between 4.86 10^9 and 6.67 10^9 M^{-1} s^{-1} by using a competition kinetic model, which utilized para-chlorobenzoic acid as a reference compound. Results showed the potential effectiveness of UV/H_2O_2 as drinking water remediation technologies for the removal of pharmaceutical compounds.

Lopez et al. [55] carried out experiments in a batch reactor in order to assess the effectiveness and the feasibility of UV processes for the decontamination of water polluted by intermediates (5-methyl-1,3,4-thiadiazole-2-methylthio and 5-methyl-1,3,4-thiadiazole-2-thiol. For both substrates, the results showed that:

(i) no degradation occurred when H_2O_2 alone was used;
(ii) UV and UV/H_2O_2 processes were both effective for degrading the substrates;
(iii) substrates degradation by photo-oxidation was always faster than by direct photolysis;
(iv) during direct photolysis, a lower substrate initial concentration lead to a faster and more efficient degradation.

Melatonin, *N*-acetyl-5-methoxytryptamine, is extensively used as the pharmaceutical or healthcare product to treat circadian rhythm, intercontinental flight disrhytmia (jet-lag syndrome) and insomnia [56]. Recent research found that melatonin may suppress the production of testosterone, decrease semen quality, and affect sexual activity and reproduction of animals and humans [57–59]. Degradation of melatonin in aqueous solutions has been investigated [60] and compared by different advanced oxidation processes, including ultraviolet radiation (UV) alone, ultraviolet radiation assisted by hydrogen peroxide (UV/H_2O_2), Fenton's reagents (Fe^{2+}/H_2O_2), and combination of Fenton's reagent with UV light (photo-Fenton, UV/H_2O_2/Fe^{2+}). The experiments were performed in a laboratory-scale batch photo reactor equipped with an 8W low-pressure mercury lamp. The degradation of melatonin by all these processes was found to follow pseudo-first order reaction kinetics. The degradation efficiency of melatonin was greatly enhanced by UV/H_2O_2 process. The fastest degradation of melatonin and highest Total Organic Carbon removal was achieved by UV/Fe^{2+}/H_2O_2 process among the four different processes tested.

Metronidazole is extensively used all over Europe for treating infections caused by anaerobic bacteria and protozoans [61-62]. Metronidazole and other antibacterial and anticoccidial drugs with nitroimidazole ring structure are suspected of being carcinogens and mutagens [63].

Degradation rates and removal efficiencies of metronidazole using UV, UV/H_2O_2, H_2O_2/Fe^{2+}, and UV/H_2O_2/Fe^{2+} have been studied in de-ionized water [64]. Degradation of metronidazole by UV and UV/H_2O_2 exhibited pseudo-first order reaction kinetics and applying H_2O_2/Fe^{2+}, and UV/H_2O_2/Fe^{2+} the degradation kinetics followed a second order behavior. Higher concentrations of hydrogen peroxide promoted the oxidation rate by UV/ H_2O_2. The addition of more ferrous ions enhanced the oxidation rate for the H_2O_2/Fe^{2+} and UV/H_2O_2/Fe^{2+} systems.

Kim et al. [65] realized an important investigation on effectiveness of UV-based processes for the removal of forty-one kinds pharmaceuticals in real wastewater using bench-scale experiment setup with a treatment capacity of $10m^3$/day. For UV process, a good removal was obtained for just a few pharmaceuticals such as ketoprofen, diclofenac and antipyrine. The removal efficiencies of macrolide antibiotics such as clarithromycin, erythromycin and azithromycin for UV alone process were very low, even with the introduction of considerable UV dose of 2768 mJ/cm^2. For UV/H_2O_2 process, a 90% removal efficiency was obtained for 39 pharmaceuticals at UV dose of 923 mJ/cm^2.

B) UV Radiation to Catalyze the Reaction of Omeprazole Degradation and Allow Their Determination Based on the Modification in the Fluorescent Properties

Omeprazole (OMP, Figure 1) (R,S)-(5-methoxy-2-[[(4-methoxy-3,5-dimethyl-2-pyridinyl)-methyl]sulphinyl]-1H-benzimidazole) is the first class of drug known as proton-pump inhibitors. It acts on the inhibition of gastric acid secretion in the stomach, and it is used to treat various acid-related gastrointestinal disorders. OMP is, in fact, a pro-drug, which is converted into a reactive intermediate sulfenamide at low pHs, combining with thiol groups of the H^+/K^+ ATPase in gastric parietal cells [66–68].

OMP contains a tri-coordinated sulphur atom in a pyramidal structure and therefore can exist in two different optically active forms, (*S*)- and (*R*)-omeprazole. OMP was first approved as a racemic mixture, but the (*S*) isomer was recently introduced on the market. Both enantiomers have a similar inhibitory effect on acid formation in isolated gastric glands from rabbits, but (*R*)-omeprazole is stereoselectively hydroxylated by cytochrome P450 CYP2C19 enzyme, resulting in an almost twofold increase in the plasma concentration for the (*S*)-isomer than for racemic OMP after the administration of equivalent doses.

This molecule undergoes extensive hepatic metabolism by the cytochrome P450 system (particularly CYP2C19 and CYP3A4) [66,69,70], and its main three metabolites are omeprazole sulfone, 5-hydroxyomeprazole, and omeprazole sulfide. Omeprazole sulphone and 5-hydroxyomeprazole are the major metabolites in plasma, while 5-hydroxyomeprazole is the major metabolite in urine [70–72].

OMP is used in the treatment of peptic ulcers, reflux oesophagitis, Zollinger–Ellison syndrome and the infection caused by *Helicobacter pylori* [66–68]. Although its elimination half-life from plasma is short—it has been reported to be about 0.5–3 h—its action is much longer compared to inhibition of acid secretion, allowing it to be used in single daily doses [73].

This pharmaceutical drug degrades rapidly in aqueous solutions at low pH values. Preformulation studies have shown that moisture, solvents, and acidic compounds have important effects on the stability of OMP [74]. UV radiation, various salts [75] and some metal ions [76] also accelerate its degradation. Its empirical formula is $C_{17}H_{19}N_3O_3S$, with a molecular weight of 345.417 [g mol^{-1}]. It is a lipophilic, weak base with pK_{a1} = 4.2 and pK_{a2} = 9 and will be degraded unless it is protected against acid conditions [77].

OMP has been recently found in surface waters of the Lambro River [78]. Stability studies have been performed especially on OMP and have revealed that it is acid labile and sensitive to light and heat [79], suggesting the possible formation of degradation products. Indeed, some problems have been observed in the establishment of the analysis of the pharmaceuticals and corresponding impurities [80-82].

In all manipulations involving OMP, care must be taken to exclude exposure to direct sunlight [83]. The photosensitivity of this substance has been noted, and the various Pharmacopeas [84-86] caution that it has to be adequately protected from light.

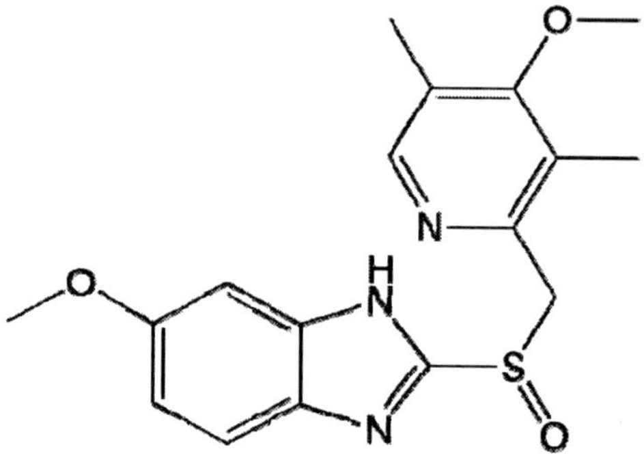

Figure 1. Chemical structure of omeprazole.

Table 2. Methods for omeprazole determination

Method	Procedure	LOD or linear range (μg mL^{-1})	Samples	Ref.
Spectrophotometry	3-Methyl-2-benzothiazolinone hydrazone following oxidation with ferric chloride	0.074	Capsules	87
	Second derivative Orthogonal function	2–42 5–35	Gastro-resistant formulations	88
	Coloured chelates of OMP in ethanol with Fe(III), Cr(III) and Co(II)	0.70, 1.98, 0.22, respectively	Dosage forms	89
Chromatography	Spectrophotometric and chromatographic techniques	WD	Pharmaceuticals	90
	HPLC	WD	Pharmaceuticals	84
Capillary electrophoresis	Micellar electrokinetic capillary chromatography	0.08– 2.0	Plasma Pharmaceuticals	91
	Non aqueous capillary electrophoresis	0.86–172.71	Pure drug	92
Spectrofluorimetry	Quenching of Tb^{3+}-1, 10-phenanthroline complex	0.016	Capsules	93
	Degradation reaction catalyzed by ultraviolet (UV) light	1.07 10^{-3}	Capsules	94

LOD: Detection limit.
WD: Without datum.

There are several analytical methods to determine OMP (Table 2) in formulations, such as spectrophotometry [87–89], chromatography [90,84], micellar electrokinetic capillary chromatography [91], and non-aqueous capillary electrophoresis [92] and spectrofluorimetry [93,94]. The official analytical method of OMP in British [85] and European Pharmacopoeias [86] is the non-selective titrimetic method with standard NaOH and potentiometric end point.

The determination of OMP by conventional UV methods suffers from interferences due to UV-absorbing compounds. Most of the reported methods for OMP determination have used UV detection systems; none of them mention the severe effect of degradation phenomenon on the methodology accuracy.

Some reported analysis of OMP solutions [84-93], mainly with UV detection systems, involve previous separative methods like HPLC, liquid-liquid extraction or other methods, lasting 15-20 minutes in each injection or extraction step for each sample and standard. The rapid decomposition of the drug, even when it is exposed to natural light, lead one to think that the measurements of UV absorbance in those methods cannot be accurate and precise since between sample/standard preparation and the instrumental detection there pass several minutes.

Because of the high sensitivity and selectivity, and being rapid and simple, spectrofluorimetric methods have been widely used for determination of a wide range of pharmaceuticals with the advantage that in most instances, derivatization is not required, and these methods are less sensitive to matrix effects than other analytical techniques [95-96].

Many factors can affect the stability of a pharmaceutical product; some of them include the stability of the active ingredient, the manufacturing process, the environmental conditions, (such as heat, light and moisture during storage), as well as some chemical reactions like oxidation, reduction, hydrolysis and racemization that might occur [97,98]. Chemical reactions occur only when a molecule is provided the necessary "activation energy". In case of photochemical reactions light provides the activation energy.

Simplistically, light is one mechanism for providing the activation energy required for many reactions.

Usually, the prediction of the photochemical behavior of a molecule is a very complex problem [99], because the photodegradation mechanism depends not only on the structure but also on experimental conditions.

According to ICH [5], the intrinsic photostability characteristics of new drug substances and products should be evaluated to demonstrate that light exposure does not result in unacceptable changes. The kinetic degradation process and the modification of OMP fluorescent properties was investigated in solution after being exposed to ultraviolet light, taking advantage of the UV decomposition of the drug for enhancing the fluorescent signal and for measuring a more stable compound.

Fluorescence Characteristics of OMP and Its Photodegradation Product

Peralta et al. [94] found that OMP, which was previously not fluorescent, produced upon UV irradiation a strongly fluorescent photoproduct with an emission maximum at 317 nm when excited at 293 nm. These wavelengths were selected to measure the fluorescence intensity. Figure 2 shows the fluorescent spectra of OMP degradation product produced by exposing the drug to direct ultraviolet light during different periods of time. The fluorescence peak intensity increases with the exposure time to UV light due to the degradation product formation. Some researchers [100–104] have proposed the scheme of OMP degradation in acid medium.

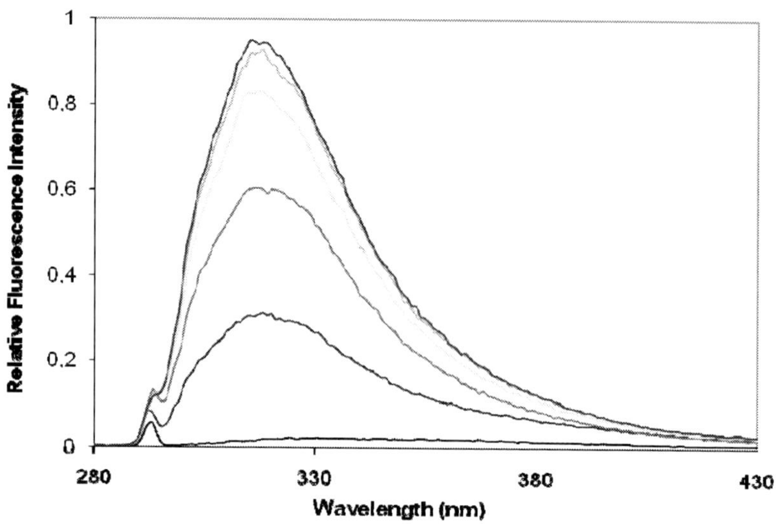

Figure 2. Emission spectra obtained for a 0.345 μg mL^{-1} of OMP methanolic solution at pH 4. From bottom to top: without exposure to UV light; Exposure time to UV light: 1 h; 2 h; 4 h; 6 h; 8 h.

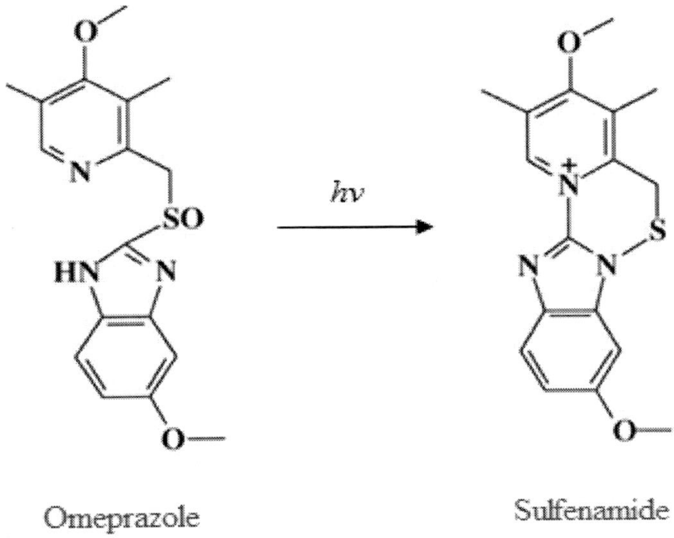

Figure 3. Proposed reaction mechanism for omeprazole.

The authors postulated a ring closure with formation of a sulfonamide derivative. The photocyclization reaction of OMP to fluorescent sulfonamide was found to be analytically useful in photochemical fluorometry. The proposed reaction mechanism is given in Figure 3.

Photoproducts (derivatives sulfonamide formed) present increased rigidity and coplanarity, resulting in an enhancement absorption coefficient and a higher fluorescence quantum yield relative to those of the analyte (OMP).

The subsequent formation of sulfenamide from OMP decomposition is in accordance with known reactions between sulfenic acids and amines. Since electron-donating groups are able to stabilize the positive charge on the sulfonam ring of the cyclic sulfonamide, the cyclic sulfonamide of OMP demonstrates highest stability [103].

Stability studies of other proton pump inhibitors, like lansoprazole, under acidic and thermal degradation conditions have been reported; only one main degradation product, the sulfonamide form, was described for acidic degradation [105–106].

This photoproduct is chemically and thermally stable on the time scale needed for analytical measurements; and the photoconversion process is very efficient, with high photochemical quantum yields.

All these facts constitute the basis of the highly sensitive and precise spectrofluorimetric method for OMP determination proposed in this work.

The effect pH on the fluorescence intensity of OMP degraded in methanolic solutions was investigated in the pH range of 2-11. The results indicated that the fluorescent intensity increases at pH 4–5, decreases between 6 and 8, and increases again at pH 9–10. The pH selected for successive measurement was 4, due to the highest fluorescence intensity obtained. Different acids were proved to adjust the pH in 1 to 5 ranges. The highest fluorescent intensity and the degradation kinetics were achieved when acetic acid was used.

A major advantage of the photochemical process owes to their mild operational conditions (ambient temperature and normal pressure).

Kinetics Determination

The kinetics of photodegradation of OMP was evaluated in methanol, adjusted at pH 4 by the addition of acetic acid 1 mol L^{-1}; the UV lamp Sankyo Denky (Japan) was used for the degradation procedure. The solutions of OMP in 10 mL graduated polyethylene centrifuge tubes were exposed to the UV irradiation for time intervals of ten min, during a period of 480 min at room temperature. The samples were placed horizontally to provide the maximum exposure area to the light source. Three samples were analyzed for each time interval. After the required time, aliquots of 4 ml were transferred to a quartz cell of a Shimadzu RF-5301PC spectrofluorimeter, for the determinations.

In the present work, the main degradation factor investigated was light. In the kinetics determination of the photodegradation, it was found that the decomposition rate is high, and OMP can be totally degraded in approximately six h at room temperature. The plot of fluorescent signal versus time is shown in Figure 4.

The velocity constant (*k*) of degradation reaction catalyzed by UV light was obtained according to Eqn. 1.

$$F = kt \qquad \text{Eqn. 1}$$

where F is fluorescent signal at 317 nm (λ_{exc} = 293 nm) at time t.

Through the evaluation of the correlation coefficients (0.9963), it can be concluded that the photodegradation of OMP in methanolic solution shows zero-order kinetics under the experimental conditions applied. The calculated zero-order degradation rate constant was k = 2.851 min^{-1}. Therefore, the exposure time of OMP to UV radiation has an important effect on all quantification measurements, especially on those in which UV light is used as the detection system.

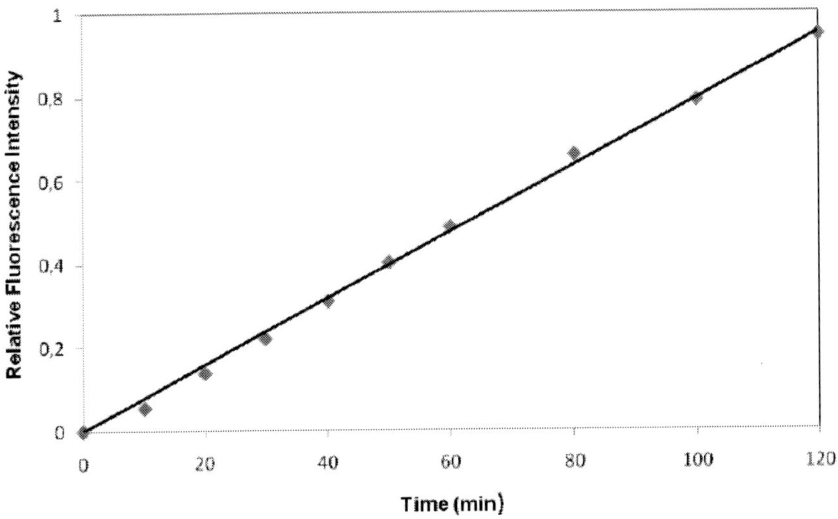

Figure 4. Representation of the equation F = kt corresponding to the determination of the order of reaction of the omeprazole photodegradation.

Methanolic solutions exposed to UV light showed a fast degradation. The dark control did not show any degradation product (the absorption UV and fluorescent spectra of OMP do not vary) under the same conditions. This indicates that the effect of heat or solvent were not responsible for the observed degradation; only the radiation causes the new product formation.

OMP photodegradation was monitored by UV–Vis spectrophotometry and spectrofluorimetry. It is evident that there was a difference between the two methodologies at the moment of monitoring the degradation process of OMP.

Data obtained by UV–Vis spectrophotometry showed that there are differences between the absorbance values before and after exposure of the drug to UV light when measured every ten min. This decomposition is shown by a shift of the absorption maximum to shorter wavelengths and, simultaneously, a decrease of absorption signal intensity [104]. However, it can be observed that this change in absorbances is small compared to the high increase in fluorescent emission signal for the same samples and exposure times. For instance, for one h of exposure time, the fluorescence emission is approximately tenfold that of non-degraded OMP.

Determination of OMP in Pharmaceutical Preparations

The proposed method was also used to determine OMP in pharmaceutical preparations (capsules). Additives and excipients did not interfere. The quantification of OMP in pharmaceuticals was carried out from the calibration graph obtained for the standard solutions. The equation for the calibration graphs was: $F = h + mC$, where F is the fluorescence intensity and C is the concentration of OMP. The assay results were in good agreement with the labeled content.

The recoveries were determined by adding various amounts of OMP to each pharmaceutical preparation and subtracting the results obtained for pharmaceuticals to which no OMP had been added. Each sample was analyzed six times; the recovery percentages obtained were in the range 94.57 to 109.66 %, with R.S.D. between 2.3 % and 4.9%.

In this work, samples and standards are simultaneously treated during 60 minutes under UV light and then the fluorescence signals corresponding to a more stable compound are immediately measured. The whole process lasts approximately 80 minutes, competing, from this point of view, with the majority of the methods described in literature. This method does not need a previous separation step since the degradation conditions and fluorescence emission parameters are almost exclusive to OMP, giving a highly selective and sensitive analytical method. In addition, the method is rather simple, with the use of few reactives and inexpensive equipment. It constitutes a worthy alternative to other costly, time consuming and expensive methods.

Conclusion

For a number of years, photochemical fluorimetry has demonstrated itself to be a very selective and sensitive method of analysis of a variety of organic compounds at low concentration levels and has been applied in pharmaceutical, clinical and environmental analyses [23,107,108].

It has developed into a versatile and simple technique, adaptable to various media such as HPLC, solid surfaces, and liquid solutions.

Several photoreactions have already been shown to produce fluorescent photoproducts and, therefore, to be analytically useful. It seems likely that many more photoreactions and photoreactive analytes will be discovered, leading to enhanced fluorescence sensitivities. Sensitized photoreactions should be good candidates from this standpoint. Also, the use of solid-state photochemical reactions has been very limited until now, and it should be developed in the future.

Because of the simplicity, sensitivity, and selectivity of photochemical fluorescence analysis, applications will no doubt continue to proliferate in the future, especially in the pharmaceutical and clinical fields.

ACKNOWLEDGMENTS

The authors wish to thank to Instituto de Química San Luis – Consejo Nacional de Investigaciones Científicas y Tecnológicas (INQUISAL-CONICET), FONCYT (Fondo Nacional de Ciencia y Tecnología), National University of San Luis (Proyect22/Q828) for the financial support.

REFERENCES

[1] Albini A; Fasani E. *Photochemistry and Photolability*, Royal Soc. Chemistry, Cambridge, 1998, p. 66.

[2] Tønnesen HH. *Photostability of Drugs and Drugs Formulations*, 2nd ed. CRC Press, Orlando, FL, 2004.

[3] Greenhill JV; McLelland MA. *Prog. Med. Chem.* 1990, 27, 51.

[4] Vadas EB. In: Gennaro AR. *Remington: The Science and Practice of Pharmacy,* 20th ed., Lippincott Williams and Wilkins, Philadelphia, 2000, pp. 980–985.

[5] ICH International Conference on Harmonization, Stability testing: *photostability testing of new drug substances and products*, 1996.

[6] Fasani E; Albini A; Gemme S. *Int. J. Pharm.* 2008, 352, 197.

[7] Dedola G; Fasani E; Albini A. *J. Photochem. Photobiol. A: Chem.* 1999, 123, 47.

[8] Garcia CV; Nudelman NS; Steppe M; Schapoval EES. *J. Pharm. Biomed. Anal.* 2008, 46, 88.

[9] Onoue S; Kawamura K; Igarashi N; Zhou Y; Fujikawa M; Yamada H; Tsuda Y; Seto Y; Yamada S. *J. Pharm. Biomed. Anal.* 2008, 47, 967.

[10] Caffieri S; Dall'Acqua S; Castagliuolo I; Brun P; Miolo G. *J. Pharm. Biomed. Anal.* 2008, 47, 771.

[11] Udenfriend S. *Fluorescence Assay in Biology and Medicine*, Vol. 2. Academic Press, New York, 1969.

[12] Fink W; Koeffier WR. *Anal. Chem.* 1970, 42, 990.
[13] Fitzgerald JM. In: Fitzgerald JM. *Analytical Photochemistry and Photochemical Analysis,* Marcel Dekker, New York, 1971, p. 145.
[14] Lukasiewicz RJ; Fitzgerald JM. *Anal. Chem.* 1973, 45, 511.
[15] Lukasiewicz RJ; Fitzgerald JM. *Appl. Spectrosc.* 1974, 28, 151.
[16] Aaron JJ; Villafranca JE; White VR; Fitzgerald JM. *Appl. Spectrosc.* 1976, 30, 159.
[17] Aaron JJ. *Methods Enzymol.* 1982, 67 F, 140.
[18] Birks JW; Frei RW. *Trends Anal. Chem.* 1982, 1, 361.
[19] Gandelman MS; Birks JW. *Anal. Chem.* 1982, 54, 2131.
[20] Gandelman MS; Birks JW; Brinkman UATh; Frei RW. *J. Chromatogr.* 1983, 282, 193.
[21] Gandelman MS; Birks JW. *Anal. Chim. Acta* 1983, 155, 159.
[22] Traore S; Aaron JJ. *Anal. Lett.* 1987, 20, 1995.
[23] Aaron JJ. In: Schulman SG. Molecular Luminiscence Spectroscopy. *Methods and Applications*. Part 3, Wiley, New York, 1993, 85.
[24] Fidanza J; Aaron JJ. *Analusis* 1981, 9, 118.
[25] Aaron JJ, Fidanza J. *Talanta* 1982, 29, 383.
[26] Aaron JJ; Ndiaye SA; Fidanza J. *Analusis* 1982, 10, 433.
[27] Tsuchiya M; Torres E; Aaron JJ; Winerfordner JD. *Anal. Lett.* 1984, 17, 1831.
[28] Kikuta C; Schmid R. J. *Phannacol. Biomed. Anal.* 1989, 7, 329.
[29] Harman PJ; Blackman GL; Philipou G. *J. Chromatogr.* 1981, 225, 131.
[30] Thsuchiya M; Aaron JJ; Torres E; Winefordner JD. *Anal. Lett.* 1985, 18, 1647.
[31] Kan RO. *Organic Photochemistry*, McGraw Hill, New York, 1966, 219.
[32] Brown RR; Bam R; Craig Jordán V. *J. Chromatogr. Biomed. Appl.* 1983, 272, 351.
[33] Aaron JJ; Diop A. *Analusis* 1985, 13, 40.
[34] Fidanza J; Aaron JJ. *J. Pharm. Biomed. Anal.* 1987, 5, 619.
[35] White VR; Fring CS; Villafranca JE; Fitzgerald JM. *Anal. Chem.* 1976, 48, 314.
[36] Fricoteaux R; Quaglia M; Aaron JJ. *4th International Symposium on Quantitative Lioninescence Spectrometry in Biomedical Sciencies.* Ghent, Belgium, May 1991.
[37] Procopio JR; Hernández PH; Hernández LH. *Analyst* 1987, 112, 79.
[38] Lefevre MF; Frei RW; Scholten AHMT; Brinkman UATh. *Chromatographia* 1982, 15, 459.

[39] Brinkman UATh; Welling PLM; De Vries G; Scholten AHMT; Frei RW. *J. Chromatogr.* 1981, 217, 463.
[40] Scholten AHMT; Brinkman UATh; Frei RW. *Anal. Chim. Acta* 1980, 114, 137.
[41] Scholten AHMT; Welling PLM; Brinkman UATh; Frei RW. *J. Chromatogr.* 1980, 199, 239.
[42] Chen D; Ríos A; Luque del Castro MD; Valcárcel M. *Analyst* 1991, 116, 171.
[43] Laasis B; Aaron JJ; Mahedero MC. *Anal. Chim. Acta* 1994, 290, 27.
[44] Laasis B; Aaron JJ; Mahedero MC. *Talanta* 1994, 41, 1985.
[45] Maafí M; Laasis B; Aaron JJ; Mahedero MC; Muñoz de la Peña A; Salinas F. *J. of lncl. Phen. and Mol. Recognition in Chem.* 1995, 22, 235.
[46] Uihlein M; Schwab E. *Chromatographia* 1982, 15, 140.
[47] Arakawa Y; Imai K; Tamura Z. *Anal. Biochem.* 1983, 132, 389.
[48] Schmid RW; Wof C; Kupferschmidt R. *13th Symposium on Column Liquid Chromatography,* Stockholm, Sweden, June, 1989.
[49] Wolf C; Schmid RW. *2nd International Symposium on Pharmaceutical and Biomedical Analysis,* York, England, April, 1990.
[50] Buser H; Müller MD; Theobald N. *Environ. Sci. Technol.* 1988, 32, 188.
[51] Halling-Sørensen B; Nors Nielsen S; Lanzky PF; Ingerslev F; Holten Lützhøft HC; Jorgensen SE. *Chemosphere* 1998, 36, 357.
[52] Heberer T. *Fresenius Environ. Bull.* 1997, 6, 438.
[53] Ternes TA. *Water Res.* 1998, 32, 3245.
[54] Yuan F; Hu C; Hu X; Qu J; Yang M. *Water Res.* 2009, 43, 1766.
[55] Lopez A; Bozzi A; Mascolo G; Kiwi J. *J. Photochem. Photobiol. A: Chem.* 2003, 156, 121.
[56] Nowak JZ; Zawilska JB. *Pharm.World Sic.* 1998, 20, 18.
[57] Yilmaz B; Kutlu S; Mogulkoc R; Canpolat S; Sandal S; Tarakci B; Kelestimur H. *Neuroendocrinol. Lett.* 2000, 21, 301.
[58] Luboshitzky R; Shen-Orr Z; Nave R; Lavi S; Lavie P. *J. Androl.* 2002, 23, 572.
[59] Tanyildizi S; Bozkurt T; Ciftci O; Sakin F. Turk. J. *Vet. Anim. Sci.* 2006, 30, 89.
[60] Xu XR; Li XY; Li XZ; Li HB. *Sep. Purif. Technol.* 2009, 68, 261.
[61] Tally FP; Sullivan CE. *Pharmacotherapy* 1981, 1, 28.
[62] Lau AH; Lam NP; Piscitelli SC; Wilkes L; Danziger LH. *Clin. Pharmacokinet.* 1992, 25, 328.
[63] Daeseleire E; Ruyck H; Van Renterghem R. *Analyst* 2000, 125, 1533.
[64] Shemer H; Kaçar Kunukcu Y; Linden KG. *Chemosphere* 2006, 63, 269.

[65] Kim I; Yamashita N; Tanaka H. *J. Hazard. Mater.* 2009, 166, 1134.
[66] Hardman JG, Limberd LE. Goodman Gilman's, *The Pharmacological Basis of Therapeutics,* 10th Ed, McGraw-Hill: New York, 2001.
[67] Gennaro AR. *Remington Farmacia,* Tomo 2, 20th Ed, Editorial Médica Panamericana: Buenos Aires, 2000.
[68] Hatlebakk JG; Berstad A. *Clin. Pharmacokinet.* 1996, 31, 386.
[69] Clissold SP; Campoli-Richards DM. *Drugs* 1986, 32, 15.
[70] Howden CW. *Clin. Pharmacokinet.* 1991, 20, 38.
[71] Renberg L; Simonsson R; Hoffman KJ. *Drug Metab. Disp.* 1989, 17, 69.
[72] Regardh CG; Andersson T; Lagerstrom PO; Lundborg P; Skanberg I. *Ther. Drug Monit.* 1990, 12, 163.
[73] Espinosa Bosch M; Ruiz Sanchez AJ; Sanchez Rojas F; Bosch Ojeda C. *J. Pharm. Biomed. Anal.* 2007, 44, 831.
[74] Davidson AG; Mc Callum A. *Drug Dev. Ind. Pharm.* 1996, 22, 1173.
[75] Ekpe A; Jacobsen T. *Drug Dev. Ind. Pharm.* 1999, 25, 1057.
[76] Hamdam II. *Pharmazie* 2001, 56, 785.
[77] Maton PN; *New Engl. J. Med.* 1991, 324, 965.
[78] Calamari D; Zuccato E; Castiglioni S; Bagnati R; Fanelli R. *Environ. Sci. Technol.* 2003, 37, 1241.
[79] Wallmark B; Lindberg P. *Pharmacology* 1987, 1, 158.
[80] Castro D; Moreno MA; Torrado S; Lastres JL. *J. Pharm. Biomed. Anal.* 1999, 21, 291.
[81] El-Kousy MN; Bebawy LL. *J. AOAC Int.* 1999, 82, 599.
[82] Karljikovic-Rajic K; Novovic D; Marinkovic V; Agbaba D. *J. Pharm. Biomed. Anal.* 2003, 32, 1019.
[83] Yuen KH; Choy WP; Tan HY; Wong JW; Yap SP. *J. Pharm. Biomed. Anal.* 2001, 24, 715.
[84] USP 29, NF 24 *United States Pharmacopeia and the National Formulary,* 2006, pp. 1758–1763.
[85] British Pharmacopoeia, 3rd Ed, *The Stationery Office:* London, 2001, pp. 1211–1213.
[86] *European Pharmacopoeia Council of Europe*, 5th Ed, 2005, pp. 3251–3253.
[87] Sastry CSP; Naidu PY; Murty SSN. *Talanta* 1997, 44, 1211.
[88] Wahbi AA; Abdel-Razak O; Gazy AA; Mahgoub H; Moneeb MS. *J. Pharm. Biomed. Anal.* 2002, 30, 1133.
[89] Salama F; El-Abasawy N; Abdel Razeq SA; Ismail MMF; Fouad MM. *J. Pharm. Biomed. Anal.* 2003, 33, 411.

[90] Gallardo V; López-Viota M; Sierra J; Ruiz MA. *Pharm. Dev. Technol.* 2009, 14, 516.

[91] Pérez-Ruiz T; Martinez-Lozano C; Sanz A; Bravo E; Galera R. *J. Pharm. Biomed. Anal.* 2006, 42, 100.

[92] Olsson J; Stegander F; Marlin N; Wan H; Blomberg LG. *J. Chromatogr.* A 2006, 1129, 291.

[93] Shaghaghi M; Manzoori JL; Jouyban A. *DARU* 2008, 16, 256.

[94] Peralta CM, Fernández LP, Masi AN, (www.drugtestinganalysis.com) DOI 10.1002/dta.243.

[95] Han Y; Wu X; Yang J; Sun S. *J. Pharm. Biomed. Anal.* 2005, 38, 528.

[96] Tong C; Hu Z; Liu W. *J. Agric. Food Chem.* 2005, 53, 6207.

[97] Nudelman NS. *Estabilidad de Medicamentos*, El Ateneo, Buenos Aires, 1975.

[98] Kommanaboyina B; Rhodes CT. *Drug Dev. Ind. Pharm.* 1999, 25, 857.

[99] Horspool WH; Song PS. *Handbook of Organic Photochemistry and Photobiology,* CRC Press, Boca Raton, FL, USA, 1995.

[100] Lindberg P; Nordberg P; Alminger T; Brändström A; Wallmark B. *J. Med. Chem.* 1986, 29, 1328.

[101] Brändström A; Lindberg P; Bergman NA; Alminger T; Anker K; Junggren U; Lamm B; Nordberg P; Erickson M; Grundevik I; Hagin I; Hoffmann K; Johnss S; Larsson S; Löfberg I; Ohlson K; Persson B; Skänberg I; Tekenbergs-Hjelte L. *Acta Chem. Scand.* 1989, 43, 536.

[102] Qaisi AM; Tutunji MF; Tutunji LF. *J. Pharm. Sci.* 2006, 95, 384.

[103] Tutunji MF; Qaisi AM; El-Eswed B; Tutunji LF. *Int. J. Pharm.* 2006, 323, 110.

[104] Riedel A; Leopold CS. *Pharmazie* 2005, 60, 126.

[105] Mathew M; Das Gupta V; Bailey R. *Drug Dev. Ind. Pharm.* 1995, 21, 965.

[106] Radi A. *Microchem. J.* 2002, 73, 349.

[107] Aaron JJ; Coly A. *Analysis Eur. JAC* 2000, 28, 699.

[108] Icardo MC; Calatayud JM. *Crit. Rev. Anal. Chem.* 2008, 38, 118.

In: Radiation Exposure
Editors: D. Balenovic and E. Stimac

ISBN 978-1-62100-869-9

Chapter 7

ULTRAVIOLET EXPOSURE MEASUREMENTS ON A CORAL REEF BY APPLICATION OF POLYPHENYLENE OXIDE FILM DOSIMETERS

***Nathan J. Downs*[*1,2], Peter W. Schouten*[2], and *Alfio V. Parisi*[1]**

[1]Department of Biological and Physical Sciences,
University of Southern Queensland, Toowoomba, Australia
[2]School of Engineering, Griffith University,
Gold Coast, Australia

ABSTRACT

The terrestrially available UVB was measured underwater on a fringing coral reef at the southern end of Australia's Great Barrier Reef marine park (25°S, 153°E) using poly 2,6-dimethyl-1,4-phenylene oxide (PPO) film dosimeters. These dosimeters have been employed in previous research applications to measure long term terrestrial UV exposures on land and in controlled aquatic environments (typically for exposures lasting several days). The value of these dosimeters is discussed in reference to their use in an underwater coral reef environment extending over periods of several weeks. The UVB measured in the autumn at the test reef site indicates that exposures in a shallow coral reef located at a sub-tropical latitude vary from between 34

---

[*] Email: downsn@usq.edu.au ; ph (+617) 46312727 ; fax (+617) 46312721

kJm^{-2} and 50 kJm^{-2} over a period of 28 days equating to a mean daily UVB exposure of between 1 and 2 kJm^{-2}. These measurements made at an approximate depth of 2 m represent approximately 5% of the daily surface UVB. The results indicate that the film dosimeters used have the potential to be applied in a number of different underwater locations to investigate the impact of solar UV radiation on exposed coral surfaces.

INTRODUCTION

The health of corals living within the Great Barrier Reef and other reef environments around the world is seen by many researchers as an indicator of global climate change particularly in regards to sea surface temperature and its subsequent effect on coral bleaching episodes [1, 2, 3, 4]. Coral bleaching, affecting the appearance and pigmentation of hermatypic corals results primarily from the loss of symbiotic zooxanthellae algae responsible for the photosynthetic generation of food products taken in by coral tissue. Currently, the health of living coral reefs worldwide as judged by the frequency of coral bleaching appears to be declining with the migration and adaptation of various coral species being unlikely to account for the sustainability of healthy coral reef environments [5].

The health of a living coral reef is dependent upon a number of factors, namely, water temperature, the absence of bacterial infection or biological predation, minimal excess dissolved and suspended nutrient concentration, and levels of ultraviolet (UV) and photosynthetically active solar radiation (PAR). Photoinhibition and bleaching evident on coral surfaces that receive a high solar exposure caused by increases in the UV and PAR irradiance [6, 7] have the potential to disturb the symbiotic balance between living coral and zooxanthellae particularly under high water temperatures when the number of zooxanthellae available to the host coral are low [5] or suffer from a reduced photosynthetic rate [8]. Furthermore, the synergistic effect of temperature and UV exposure [9] is likely to affect coral health due to the stratification of warmer sea water resulting in calmer clearer conditions and greater UVB (280 nm to 320 nm) penetration depths [10]. Undoubtedly increased levels of UV radiation due in part and in association with high water temperature, has a harmful effect on coral health. The causative role of solar UV radiation influencing the mutagenesis of DNA in the living tissue of plants and animals is likely to be of additional consequence, particularly in regards to the photosynthetic efficiency of corals [11].

MATERIALS AND METHODS

PPO Dosimeter Film Calibration

The UV responsive poly 2,6-dimethyl-1,4-phenylene oxide (PPO) film dosimeter was calibrated for use on the fringing coral reef environment to measure solar UVB exposures over an extended time period. The dosimeter is suited to extended periods of solar exposure in-air [12] and has been shown to be effective at measuring UVB exposures extending over several days duration in underwater environments [13]. The physical dosimeter employed in this research application is comprised of the chemical PPO cast at the University of Southern Queensland in thin film form to a thickness of approximately 40 μm. The PPO film is placed into a polyvinyl chloride (PVC) holder of size 3.3 cm x 3.3 cm with an aperture of 1.2 cm x 1.6 cm. The film is secured to the holder with tape and removed after measurement to enable reuse of the dosimeter frame. The change in optical absorbance caused by UV exposure upon the PPO film is measured by a spectrophotometer (model 1601, Shimadzu Co., Kyoto, Japan) at 320 nm. This wavelength lies in the UVB range and has been determined to show a measurably significant change in optical absorbance in PPO [13]. The accuracy threshold for optical absorbance measurements of the spectrophotometer utilised in this research application has been quoted as ± 0.002 by the manufacturer. After each exposure to solar UV, the discrepancy in optical absorbance at 320 nm (ΔA_{320}) for each PPO dosimeter used in the measurement campaign was measured, where ΔA_{320} is calculated using the equation:

$$\Delta A_{320} = A_{320}^{final} - A_{320}^{initial}$$

where, A_{320}^{final} is the post-exposure optical absorbance measurement taken at the wavelength of maximum change in optical absorbance of the film at 320 nm after solar exposure in the marine environment and $A_{320}^{initial}$ is the film absorbance measurement taken at the same wavelength before solar exposure [13]. At the conclusion of the underwater exposure period and before measurement in the spectrophotometer, any particulate matter remaining on the PPO film surface was removed with a blast of distilled water. Cleaned PPO film was left to dry in a darkened container prior to post-exposure measurement.

To increase the overall accuracy of the measurements, the ΔA_{320} value was measured over four positions across the film surface with the mean of these values used to calculate the UV exposure by application of an exposure-response calibration function. The exposure-response calibration for the PPO in seawater employed for this research campaign is given in Figure 1.

The UVB calibration function given in the figure was calibrated to the measured solar UVB irradiance by a StellarNet EPP2000 spectrometer (StellarNet EPP2000 C-UV-VIS, Tampa, Florida) within a UVB inclusive waveband running from 298 nm to 320 nm. The lower limit of 298 nm was chosen as it was the cut-off wavelength at which the EPP2000 spectrometer could obtain reliable underwater data. The upper limit of 320 nm was chosen as it has been used by photobiologists to denote the end of the UVB waveband and is more applicable to this research. The EPP2000 spectrometer consists of a sensitive charge coupled device (CCD) working in unison with a high speed parallel digitiser interface with a large spectral scan memory. This setup allowed for almost instantaneous measurement of incoming solar spectra over the UVB waveband. Input was provided by a SMA 905 fibre optic connector with a small diffuser attached to the end. Control of the EPP2000 and spectral data collection was performed by a laptop computer. The PPO film dosimeters were calibrated in sea water tanks located in proximity to this instrument.

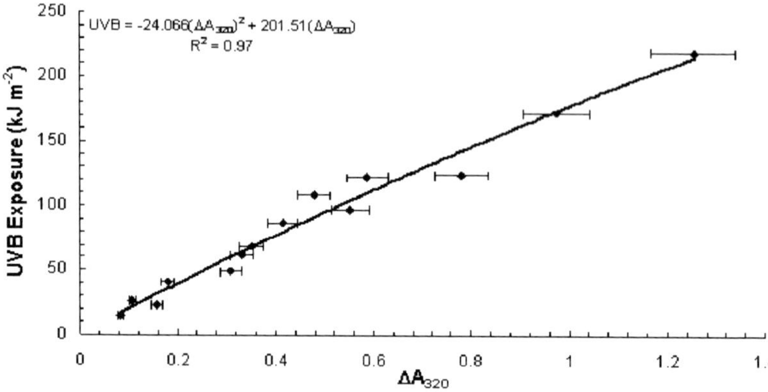

Figure 1. Exposure-response calibration curve for PPO film dosimeters measured in seawater. Calibration measurements were made in sea water at the USQ Toowoomba campus between February 2007 and April 2007. The error bars represent the ±9% uncertainty present in each underwater PPO film calibration measurement (Schouten et al. 2007).

As seen in Figure 1, the uncertainty in calibrated PPO film dosimeters increases with exposure period. For dosimeters placed in the reef environment, ΔA_{320} was measured below 0.5.

RESULTS

Measurement of UV Exposures on the Coral Reef

PPO ultraviolet film dosimeters were placed in proximity to living coral colonies located in shallow water south of the Great Barrier Reef marine park area in Hervey Bay, Australia (25°S). A total of 8 dosimeters were placed at low tide in two groups of four, each group being attached to a brick weight and placed at the same depths at the selected reef site for a period of four weeks between 28 March and 25 April 2008. An image of the PPO dosimeters placed in proximity to a *Dendrophylliidae* coral (*Turbinaria reniformis*) at the measurement site is shown in Figure 2. The diurnal tidal water level was found to vary at the reef site between 0.4 m to 3.8 m [14] above the placement depth of the film dosimeters.

The underwater UVB exposure was measured on a horizontal surface by the PPO film dosimeters by placement on the sea bed located adjacent to two *Turbinaria reniformis* colony sites which were separated by approximately 50 m. Sites were selected to be at the same approximate depth.

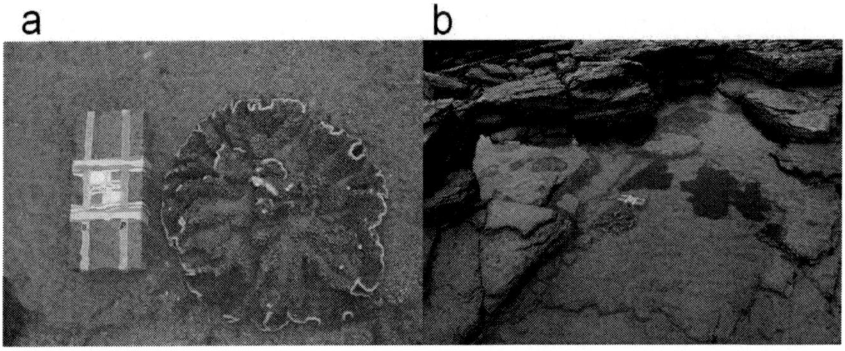

Figure 2. PPO film dosimeter placement. (a) The film dosimeters were attached to a brick and placed in proximity to Turbinaria reniformis coral colonies. (b) A wider view of the placement site pictured in (a) showing the presence of other coral colonies in the fringing coral reef environment during placement at low tide.

The mean calibrated exposure at each of the two coral reef placement sites was found to be 50 $\pm$ 9 kJm^{-2} (1σ) and 34 $\pm$ 21 kJm^{-2} (1σ). The reef exposures measured over 28 days were recorded during a period of frequent high cloud coverage. The measured underwater exposures represent approximately 5% of the surface UVB likely to be experienced in the reef exposure period. This estimate is based on an average daily UVB exposure of 29 kJm^{-2}, which is the average daily exposure measured previously at a site located in Southern Queensland at a similar latitude to the reef location [15]. The underwater exposures measured in this research approach the limits of previously published calibration functions for underwater UVB polysulphone film (PSF) dosimeters of 1.5 to 40 kJm^{-2} determined by Dunne [16]. The advantage of PPO film dosimeters is their extended dynamic range, making PPO film suitable for longer term exposure regimes.

Spectral Variations in the UV Waveband

The penetration of UVA solar radiation was measured to be slightly stronger than the transmission of solar UVB radiation by sampling water from the test reef location. Water samples obtained from the reef measurement location were analyzed using a spectrophotometer (model 1601, Shimadzu Co., Kyoto, Japan) across the entire terrestrial UV waveband from 280 nm to 400 nm. The transmission spectrum of water sampled from the reef site was measured at 86% at 280 nm increasing to 90% at 400 nm. Similarly the absorption of UV was slightly less in the longer UV wavelengths, varying from between 0.06 to 0.04 at 280 nm and 400 nm respectively. The more abundant proportion of UVA available at the Earth's surface coupled with the increased penetration of UVA into the sea water measured at the study site is likely to present marine organisms including fringing coral reef structures with a biologically significant spectral region which may play an important role in coral bleaching episodes. This is an avenue for future research.

PPO Film as an Underwater UVA Dosimeter

Any changes in UV radiation reaching the surface of the Earth due to changing atmospheric chemistry, particularly stratospheric ozone concentrations, in addition to changing sea surface levels and water temperatures are likely to elicit a change in underwater UVA and UVB exposure incident upon coral reef systems. Future research applying PPO film dosimeters are likely to benefit from measurement in both the UVA and UVB. UVB blocking mylar films have been added as filters to phenothiazine film dosimeters to measure the UVA in terrestrial environments [17]. As PPO has been found to be slightly sensitive to UVA [18], this technique has the potential to be transferred to PPO films deployed in an underwater environment making cost effective simultaneous UVA and UVB measurements possible.

Conclusion

The potential influence of solar UVA affecting the health of zooxanthellae living in stressed coral tissues may have a more biologically significant impact on the bleaching response of some corals, being more abundant in the underwater environment than the UVB. Measurements taken using PPO film dosimeters in the UVB waveband do however give a reasonable cost effective first approximation of the UV affecting exposed coral surfaces. To date there is limited information available on the solar spectral response of corals and their associated zooxanthellae algae. PPO film dosimeters provide a convenient method that can be used to measure the UVB exposure received in the natural coral reef environment and may be applied in future studies to better quantify the influence of UVB radiation in bleached coral environments or to monitor the long term underwater radiation received by coral surfaces.

The calibrated UVB exposures recorded at the test reef site and measured in proximity to living coral colonies at this location give a first approximation of the UV exposure received by fringing reef corals which experience daily sediment loading in a turbid tidal system. The exposures presented here were found to be approximately 5% of the ambient UVB measured over a four week period. Previous research [19] has determined the 1% UVB depth for an inshore reef to be 3 m using a semi-submersible scanning spectroradiometer. The tidal depth above the PPO dosimeter placement sites in this research varied from between 0.4 m to 3.8 m. The low cost and extended dynamic range of PPO film makes the use of dosimeters an attractive option for studies involving simultaneous measurements of UVB in a natural coral reef environment. Application of PPO film dosimeters deployed in high density to measure the UV exposure received by the zooxanthellae associated with many hermatypic corals provides an additional method that may be used to directly measure UVB exposure to coral surfaces negating the need to employ spectral instrumentation.

ACKNOWLEDGMENTS

The authors would like to thank Oliver Kinder from the Faculty of Sciences, USQ for assistance with manufacturing necessary reef dosimeter equipment. Appreciation is also given to research associate, Liam Ritchie for assisting with the coral reef dosimeter placement and recovery.

REFERENCES

[1] T.J. Goreau, R.L. Hayes. (1994). Coral bleaching and ocean "hot spots", *Ambio Vol.* 23, pp 176-180.

[2] S. Anderson, R. Zepp, J. Machula, D. Santavy, L. Hansen, E. Mueller. (2001). Indicators of UV exposure in coral and their relevance to global climate change and coral bleaching, *Hum. Ecol. Risk Assess*. Vol. 7, pp 1271-1282.

[3] T.P. Hughes, A.H. Baird, D.R. Bellwood, M. Card, S.R. Connolly, C. Folke, R. Grosberg, O. Hoegh-Guldberg, J.B.C. Jackson, J. Kleypas, J.M. Lough, P. Marshall, M. Nyström, S.R. Palumbi, J.M. Pandolfi, B. Rosen, J. Roughgarden. (2003). Climate Change, human impacts, and the resilience of coral reefs, *Science* Vol. 301, pp 929-933.

[4] J.M. Lough. (2008). 10th Anniversary Review: a changing climate for coral reefs, *J. Environ. Monit.* Vol. 10, pp 21-29.

[5] O. Hoegh-Guldberg. (1999). Climate change, coral bleaching, and the future of the world's coral reefs, *Marine and Freshwater Research* Vol. 50, pp 839-866.

[6] D.F. Gleason, G.M. Wellington. (1993). Ultraviolet radiation and coral bleaching, *Nature* Vol. 365, pp 836-838.

[7] B.E. Brown, R.P. Dunne, M.S. Goodson, A.E. Douglas. (2000). Bleaching patterns in reef corals, *Nature* Vol. 404, pp 142-143.

[8] M.E. Warner, W.K. Fitt, G.W. Schmidt. (1996). The effects of elevated temperature on the photosynthetic efficiency of zooxanthellae in hospite from four different species of reef coral: a novel approach, *Plant Cell Environ.* Vol. 19 pp 291-299.

[9] J.M. Shick, M.P. Lesser, P.L. Jokiel. (1996). Effects of ultraviolet radiation on corals and other coral reef organisms, *Global Change Biol.* Vol. 2 pp 527-545.

[10] D.A. Siegel, A.F. Michaels. (1996). Quantification of non-algal light attenuation in the Sargasso Sea: implications for biogeochemistry and remote sensing, *Deep-Sea Res.* Vol. 43 pp 321-346.

[11] C. Ferrier-Pagès, C. Richard, D. Forcioli, D. Allemand, M. Pichon, J.M. Shick. (2007). Effects of temperature and UV radiation increases on the photosynthetic efficiency of in four Scleractinian coral species, *Biol. Bull.* Vol. 213 pp 76-87.

[12] Davis, G.H.W. Deane, D. Gordon, G.V. Howel, A.K. Ledbury. (1976). A world-wide programme for the continuous monitoring of solar ultraviolet radiation using poly (phenylene oxide) films and a consideration of results. *J. Appl. Polymer Sci.* Vol. 20 pp 1165-1174.

[13] P.W. Schouten, A.V. Parisi, D.J. Turnbull. (2007). Evaluation of a high exposure solar UV dosimeter for underwater use. *Photochem. Photobiol.* Vol. 83 pp 931-937.

[14] BOM (Bureau of Meteorology), Tide Prediction for Australia, South Pacific and Antarctica (from the National Tidal Centre), Commonwealth of Australia, Bureau of Meteorology, Accessed 22 June 2011, Available at: http://www.bom.gov.au/oceanography/tides/

[15] J. Sabburg, A. Parisi, J.C.F. Wong. (1997). Ozone, cloud, solar and UV-B levels at a low pollution, southern hemisphere, sub-tropical site for winter/spring 1995, Australas. *Phys. Eng. Sci. Med.* Vol. 20 pp 198-202.

[16] R.P. Dunne. (1999). Polysulphone film as an underwater dosimeter for solar ultraviolet-B radiation in tropical latitudes, *Mar. Ecol. Prog. Ser.* Vol. 189 pp 53-60.

[17] A.V. Parisi, M.G. Kimlin, D.J. Turnbull, J. Macaranas. (2005). Potential of phenothiazine as a thin film dosimeter for UV-A exposures. *Photochem. Photobiol. Sci.* Vol. 4 pp 907-910.

[18] Schouten, P.W., Parisi, A.V., Turnbull, D.J. (2010) Usage of the polyphenylene oxide dosimeter to measure annual solar erythemal exposures. *Photochem. Photobiol.* Vol. 86 pp 706-710.

[19] R.P. Dunne, B.E. Brown. (1996). Penetration of solar UVB radiation in shallow tropical waters and its potential biological effects on coral reefs; results from the central Indian Ocean, and Adaman Sea, *Mar. Ecol. Prog. Ser.* Vol. 144 pp 109-118.

In: Radiation Exposure
Editors: D. Balenovic and E. Stimac
ISBN 978-1-62100-869-9

Chapter 8

RADIATION EXPOSURES IN MEDICAL IMAGING

Márcio Sommer Bittencourt and Henrique Lane Staniak
Hospital Universitário – Universidade de São Paulo
São Paulo, Brazil

ABSTRACT

The last decades have seen an unprecedented increase in technology in medicine. Among this development, imaging has taken one of the leading positions. Nevertheless, this increase in medical imaging use comes at a significant cost. The most common medical imaging techniques used radiation to acquire the image. From the initial X ray technology form the end of the nineteen century to today's computed tomography and nuclear medicine technology, medical image came from an insignificant to the second most important cause of radiation exposure throughout the world.

Although this information may sound negative, medical imaging has an invaluable use in diagnostic and therapeutic medicine. Therefore, it should not be blamed for the radiation exposure; it should only be controlled and adequately used to optimize the medical benefits while reducing the risks associated with the increase in radiation exposure.

In this chapter we will discuss the increase in radiation exposure from medical imaging, while focusing on potential technical interventions that can help reducing exposure and optimize equipment to prevent the use of unnecessary radiation.

INTRODUCTION

Medical imaging is a broad field which includes all procedures used to view the human body for clinical purposes. They are used to screen, diagnose and guide treatment for virtually every disease. Not only that various image modalities help guide interventions such as biopsies, angiographies, stent placements or catheters.

Various modalities of medical imaging exist, each consisting of different technologies. Magnetic resonance imaging (MRI) uses magnetic fields to produce images, whereas ultrasound equipment uses high-frequency sound waves to evaluate body structures. However, most of the medical imaging technologies used today relies on radiation to produce images. This includes computed tomography, standard x-rays (radiography), fluoroscopy and nuclear medicine. Each of them uses a variable amount of radiation, which depends on the part of the body being examined, the duration and extension of the scan, as well as various parameters which set up the amount of energy in use. Radiography usually uses very low radiation (see table 1). Those examinations are restricted to the part of the body being evaluated. This part of the body receives the radiation which creates a two-dimensional image. Although radiography includes 74% of the imaging procedures using radiation in the U.S.A., due to their low dose, they are only responsible for 11% of the total radiation[1].

In computed tomography scans (also referred to as CT or CAT scan) the source of radiation rotates around the patient. The x-rays pass through the patient's body and a computer is used to reconstruct various cross-sectional images of the particular area under evaluation. This sequence of images can also be reconstructed in various forms to create other 2 dimensional and 3 dimensional images.

During fluoroscopy the x-rays pass through the patient's body during a period of time, creating a live moving image. This technique allows real time images and evaluation of moving parts of the body such as the heart. It usually is used during procedures such as cardiac catheterization and during surgical procedures.

Table 1. Radiation Doses from Various Types of Medical Imaging Procedures[2]

Type of Procedure	Average Adult Effective Dose (mSv)	Estimated Dose Equivalent (No. of Chest X-rays)
Dental X-ray	0.005-0.01[6a]	0.25-0.5
Chest X-ray	0.02	1
Mammography	0.4	20
CT	2-16[6b]	100-800
Nuclear Medicine	0.2-41[6c]	10-2050
Interventional Fluoroscopy	5-70[6d]	250-3500

Nuclear medicine examinations use completely different approach. In those procedures a radioactive substance, called radiotracer, is given to the patient, most commonly intravenously. This substance's radiation is then measured by a gamma camera outside the body. This allows the tracer to be located and quantified. One very well-known form of this examination is the positron emission tomography (PET) scan. In nuclear examinations the radiation is given to the patient and then measured, whereas in the other forms of image the radiation comes from the device used.

CT, nuclear medicine and fluoroscopic scans all use continuous extended and/or repeated radiation exposure. Therefore, they all expose the patient to significantly higher amounts of radiation than standard radiography (see table 1). Those 3 forms of imaging procedure are responsible for 89% of all medical imaging radiation exposure in the U.S.A.

Radiation Exposure

The total exposure to ionizing radiation has steadily increased over the last 20 years. Most of the increase is related to the aforementioned medical imaging procedures. Estimates suggest that about 67 million CT scans, 18 million nuclear medicine and 17 fluoroscopic procedures were performed in 2006[3]. Since the publication of this data there has been an increase in public concern over the radiation exposure. Various medical publications have addressed the problem, and even a few non-medical publications have already brought this concern up. The main reason for concern is that ionizing radiation has the potential to damage the DNA, increasing the future risk of cancer development. Although publications and even websites[4] try to estimate or calculate individual risk, no single study has ever been able to demonstrate that usual medical imaging procedures increase lifetime cancer risk. Most of the data available on radiation exposure and cancer risks is derived from the nuclear bombs and nuclear power plants accidents. The proportional risk was calculated based on those exposures, however, the main caveat of this data is that the amount of radiation exposure in those situations was extremely high, above 100mV, whereas medical imaging procedures most of the time do not reach more the 10% of this value. Nevertheless, due to the potential risks, imaging procedures should always try to keep their radiation levels to a minimum. Various medical societies have introduced the concept of ALARA (as low as reasonably achievable) to indicate that every procedure should be performed with the lowest radiation possible.

Accidental exposure can cause injuries, burns and hair loss. Care should always be taken when dealing with radiation. However, those accidents are rare, and most of the concern leis on the potential risk of repeated routine procedures, as the cumulative doses progressively increase with time.

Unnecessary Radiation

Every single procedure described earlier exposes the patient to radiation. Nonetheless, not all requested imaging procedures performed are necessary. When performed according to routine clinical standards and with an appropriate indication the benefits of those procedures out weight the risk by far and large.

Therefore, if proper precautions are taken, the radiation exposure and risk can safely be minimized without significant loss of quality in medical care for those patients. Unnecessary radiation might be used when a scan has an inappropriate indication, i. e. when it is unlikely that it will help patients diagnose or treatment; or when the scan's parameters set up are higher than necessary to acquire the requested image. Not only that, any given time an imaging procedure performed has a poor image quality which does not allow a correct reading, this must also be taken as unnecessary radiation.

We can split the unnecessary radiation exposure in two main groups: the issues related to the device used, and the issues related with the clinical decision making process.

Issues Related to the Device Use

A wide variation in the amount of radiation used for the same procedure performed at different sites has been demonstrated. Some data suggest up to a 13-fold increase in the radiation from one institution to another. The standardization of procedures could play a role in balancing the adequate amount of radiation versus the necessary image quality.

Since many of the medical staff and supporting team may not have been adequately trained to understand the potential risks of radiation and the technical aspects related to radiation dose used and safety procedures, the first step in adjusting the those is to provide adequate training opportunities for the whole team. The second step is to use the acquired information to redesign image acquisition protocols. This is particularly important as technology evolves, since newer equipment has better image quality, more advanced hardware and software, allowing newer dose reduction strategies which were not previously available.

One other important task is to guarantee the team performing the procedures should keep an adequate log of the radiation dose of each scan performed. This log should be analyzed frequently, and the standard doses used should be compared with the same procedure performed at other centers. The official norms for radiation dose are known as "diagnostic reference levels", which correspond to the 75th or 80th percentile of the distribution measured for a particular image procedure[5]. These values should be used as reference standards which should not be exceeded in any given procedure without a known reason. The American college of radiology, the American association of physicists in medicine and the food and drug administration (FDA) have all worked on those standards.

The appropriate way to reduce radiation exposure on each form of medical imaging is different. The particular set up for each method will not be extensively covered in this review. However, examples of those adjustments include kV and mAs reduction in CTs, reduction of tracer volume injected in nuclear medicine scans and reduction of the duration of fluoroscopic time.

Issues Related to Clinical Decision

Not only ordering physicians might not be fully aware of the amount of radiation used at a given procedure, but also they have no access to all previous procedures performed or total previous radiation the patient has previously been exposed to. This might lead physicians to order procedures that have already been performed.

Not only that, sometimes physicians might over order scans without proper indication or indicate procedures that are deemed unnecessary. To reduce that, ordering physicians should be trained on the diagnostic efficacy of the requested procedure, as well as on potential diagnostic options which do no expose patients to radiation (ordering a nuclear magnetic resonance instead of a CT scan if indication is appropriate, for example). Various medical societies have written consensus on appropriateness criteria for various imaging techniques and clinical indications.

Radiation Reduction

The FDA has created a collaborative initiative to reduce unnecessary radiation exposure from medical imaging[7]. This initiative takes steps to mitigate the unnecessary exposure from various medical imaging modalities.

- Step 1 – require manufacturers of devices to incorporate safeguards into the equipment design, labeling, and user training.
- Step 2 – Partner with performing centers to incorporate key quality assurance practices into accreditation and participation criteria for imaging facilities and hospitals
- Step 3 – Recommend that health care professionals continue efforts to develop reference levels and radiation exposure registries
- Step 4 – Establish requirements for device manufacturers to record radiation exposure in patients' medical records
- Step 5 – Recommend that healthcare professionals continue the development of appropriateness criteria for medical imaging procedures
- Step 6 – Increase patient awareness by providing patients with tools to track their personal medical imaging history.

Especial Groups

To especial populations must be highlighted as particularly high risk for radiation exposures: pregnant women and children. Both populations high a relatively small risk of radiation complications. However, their risk is significantly higher than for the adult population.

This does not change any of the general principles addressed above. The procedure indication, as well as device set up to reduce radiation to as low as reasonably possible should always be recommended. In particular, considering an alternative not associated to radiation diagnostic procedure is always advisable. For pregnant women, whenever possible the scan should be postponed until delivery. In situations when this cannot be done, the use of abdominal shields to protect the child is recommended.

Consensus data suggest a maximum radiation dose for pregnant women of 5 mSv during the whole pregnancy.

Conclusion

The use of medical imaging procedures has a great impact in clinical outcomes of patients. The vast majority of the procedures are well indicated and add significant improvement in patient quality of care. However, CT scan, nuclear medicine scans and fluoroscopic examinations pose a cumulative radiation exposure risk that should not be minimized. Quality control measures should always be used to ensure no unnecessary radiation is given to patients and to ensure that the clinical utility of the scans is high, and to ensure that the benefits of the procedure out weight any potential risks associated with radiation exposure.

References

[1] National Council on Radiation Protection and Measurements, *NCRP Report No. 160: Ionizing Radiation Exposure of the Population of the United States*, March 3, 2009, pp. 142-146.

[2] Mettler, Jr. FA, et al., July 2008.

[3] National Council on Radiation Protection and Measurements, *NCRP Report No. 160*, March 2009, pp.

[4] http://www.xrayrisk.com/
[5] NRPB/RCR, "Patient Dose Reduction in Diagnostic Radiology," *Doc. NRPB*, 1990, Vol. 1, No. 3.
[6] http://www.fda.gov/radiation-emittingproducts/radiationsafety/radiationdosereduction/ucm199904.htm.

In: Radiation Exposure
Editors: D. Balenovic and E. Stimac
ISBN 978-1-62100-869-9

Chapter 9

HOW TO USE UV RADIATION TO MONITORE THE PHOTOSTABILITY OF PHARMACEUTICAL PRODUCTS

***Clayton Anderson de Azevedo Filho*[1, 3,4*], *Daniel de Filgueiras Gomes*[2] *and* *Beate Saegesser Santos*[1,4]**

[1]Programa de Pós-Graduação em Ciências Farmacêuticas, Departamento de Ciências Farmacêuticas, UFPE, Recife, Brazil
[2]IEEE Member; Instituto Federal de Educação, Ciência e Tecnologia de Pernambuco, IFPE, Ipojuca, Brazil
[3]Associação Caruaruense de Ensino Superior, Faculdade ASCES, Caruaru, Brazil
[4]Research group on Parenteral Nutrition, Dept Pharmaceutical Sciences, UFPE, Brazil

ABSTRACT

In recent years, there has been a greater concern with the effects of ultraviolet (UV) radiation in pharmaceutical products such as drugs and

* Corresponding author at: Av. Prof. Arthur de Sá, s/n, Cidade Universitária, 50740-521 Recife, PE, Brazil. Tel.: +55 81 21268511; fax: +55 81 21268511. E-mail address: claytonazevedo@hotmail.com (C.A. de Azevedo Filho).

health materials. The photostability is an important parameter to evaluate the demanded quality, security and effectiveness for the safe use of health products. Due to environmental conditions in which the product will be exposed, such as temperature, humidity and light during its marketing period, photostability tests are important to monitor the degradation reactions and to predict shelf life of the final product in normal storage conditions. The sun is the most important source of UV radiation, although visible and infrared light constitute the major fraction of the radiation reaching Earth´s atmosphere. Artificial lights are also sources of UV radiation. Incandescent lamps, halogen lamps and fluorescent tubes are the most commonly used light sources in homes and work places. Incandescent lamps give practically no UV radiation, while halogen lamps operate at a higher temperature and give small fluence rates of UVA radiation. Fluorescent tubes contain mercury vapor which give a typical band spectrum showing two strong lines located at the UV region (λ = 254 and 313 nm) which are not completely filtered by the glass tube. Photochemical degradation can be an important factor in the stability of pharmaceuticals especially under exposure of high energy radiation such as the UV radiation. So, not only the drug photostability evaluation is an important step in this process, but also it is important to know and to quantify (i) the formed byproducts, (ii) the degradation kinetics and (iii) the influence of light on the original formulation. It is usually difficult to predict the light effect in the compounds and the consequences to the stability of the final preparation. So, to perform photostability testing, it is necessary to ensure that the radiation is homogeneously distributed on the surface area where the product will be exposed and to precisely determine the radiation dose received by the samples when placed in the exposure area. Thereby, it is necessary to design and to qualify a photostability chamber suitable to perform photostability testing and to guarantee the reliability of the results provided by it. Either a commercial or a home-built photostability chamber needs to ensure homogeneous light dose on the studied sample. There are two experimental ways to qualify the source of radiation: (i) by applying actinometer systems (chemical compounds with a linear photodegradation rate) or (ii) by using a physical device on which the number of photons applied can be fully determined. The main purpose of this Chapter is to discuss the procedures involved in the degradation induced by UV radiation and how to monitore and to determine the correct conditions for performing photostability testing in a photostability chamber.

INTRODUCTION

Stability is an important parameter to evaluate the quality, safety and efficacy required for the sanitary registration of pharmaceutical products. The determination of this parameter is based not only on compliance with legal requirements but it also reflects a concern for public health, since any instability may be related to loss of therapeutic effect or consumer exposure to the toxic effects of the product degradation. Stability tests are important (i) to monitor the degradation reactions and (ii) to predict the period of validity of the final product in normal storage conditions as well as the environmental conditions in which the product will be exposed during the marketing of the product throughout its shelf life [1-3].

Photochemical degradation is a possible feature that decreases the stability of pharmaceuticals, especially under exposure of high energy radiation such as the ultraviolet (UV) radiation. Light can have effects on the active principle in a drug formulation, as well as on the final product or packaging. This may be visually detected as bleaching of colored compounds or a discoloration of colorless products. So, photochemical stability or photostability is an important parameter to evaluate the demanded quality, security and effectiveness for the safe use of health products. In recent years, there has been a greater concern with the effects of radiation in pharmaceutical products such as drugs and health materials. Due to environmental conditions in which the product will be exposed, such as temperature, humidity and light during its marketing period, photostability tests are regularly required to monitor the degradation reactions and to predict shelf life of the final product in normal storage conditions [1, 4].

It is usually difficult to predict the light effect in the compounds and the consequences to the stability of the final preparation. Measurement of the light source exposure levels on the object is crucial in developing and maintaining protective measures to combat photodegradation [5]. It is also important to be able to correlate the amount of light exposure to (i) understand its influence on the original formulation; (ii) the byproducts formed in terms of its identification and quantification and (iii) establish the degradation kinetics. These measurements that will be discussed in the next sections comprehend what is known as photostability testing [5, 6]. Several countries have published guidelines for the evaluation of drug stability that include photostability testing. In the 1980's the European Community (now European Union) pioneered in setting up the harmonization of regulatory requirements and has undertaken to develop a single market for pharmaceuticals. The success achieved in Europe demonstrated that harmonization was feasible. At the same time, there were bilateral discussions between Europe, Japan and the U.S. on the possibilities for harmonization. It was, however, the World Health Organization (WHO) during the International Conference of Drug Regulatory Authorities (WHO International Conference of Drug Regulatory Authorities - ICDRA) in Paris in 1989 which discussed specific plans of action. Shortly afterwards, the authorities have joined the International Federation of Pharmaceutical Manufacturers and Associations (IFPMAS) to discuss a joint regulatory industry on the international harmonization and designed several guides to help the quality control of pharmaceutical products. One of these is entitled "Guidance for Industry Q1B Photostability Testing of New Drug Substances and Products" [3, 4, 6].

The regulators in Europe, Japan and the U.S. authorities and representatives of pharmaceutical companies created in 1990 an International Committee for Harmonization of Technical Requirements for Registration of Pharmaceuticals for Use in Human Beings (ICH). The objective of ICH is to increase international harmonisation of technical requirements to ensure that safe, effective, and high quality medicines are developed and registered in the most efficient and cost-effective manner. These activities have been undertaken to promote public health, prevent unnecessary duplication of clinical trials in humans, and minimize the use of animal testing without compromising safety and effectiveness. [6, 7]. The ICH guideline Q1B for photostability testing gives guidance on the basic testing protocols required to evaluate the light sensitivity and stability of new drugs and products [4].

The present Chapter deals with the main features related to photostability testing specially the proper requirements to ensure reproducibility of the experiments and the quality of the final results. First of all the fundamental terminology used in the light-matter interaction is described, followed by the main steps concerning photostability testing.

BASICS IN PHOTOSTABILITY TESTING

Drug substance and its products can be exposed to any one of a number of natural and artificial sources with different intensity of electromagnetic radiation (EMR) during the synthesis, manufacturing or storage in the warehouse manufacture, packaging, distribution, storage on pharmacy shelves and finally at the point of use. These exposures may include Hg vapor lamps, fluorescent and incandescent lighting, window-filtered sunlight and, in some cases, direct sunlight. It is also possible that the pharmaceutical products to be exposed to a combination of indoor lighting (e.g., white fluorescent tubes or tungsten–halogen) and window-glass–filtered sunlight during storage by the consumer. It is clearly difficult to predict the amount of UV and VIS irradiation to which the product is exposed during its shelf life [8, 9].

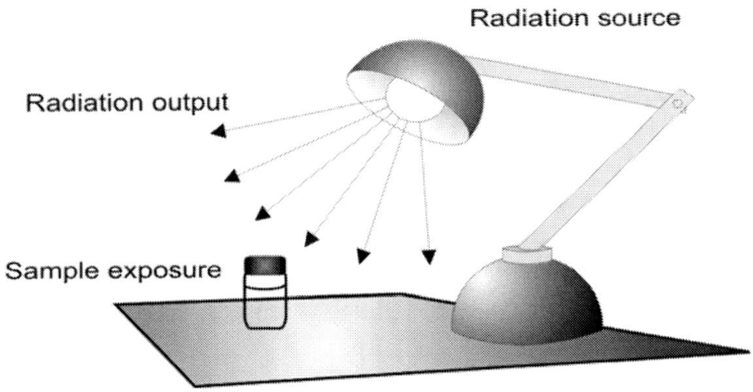

Figure 1. Schematic diagram showing the main steps related to a photostability testing: the radiation source, the radiation output and the real radiation dose exposing the sample.

To provide a similar picture that may reproduce the radiation effects to which the samples are exposed it is necessary to create a set up with controlled experimental conditions. This experimental set up is denominated as photostability testing and is required to (i) produce the same spectral conditions and (ii) to measure and control the total radiation to which the sample was exposed during a certain amount of time (Figure 1). In a photostability testing the spectral power distribution (SPD) and the overall illuminance should provide a "worst-case" exposure. Usually photostability testing is performed in chambers with controlled radiation conditions. The chambers diversify according to the design: they may be commercial or non-commercial chamber, show different positions of sample radiation exposure – horizontal and vertical – with and without rotation - and may possess different types and designs of lamps.

To perform photostability testing it is necessary to ensure that the radiation is homogeneously distributed on the exposure surface area where the product will be exposed and also to determine precisely the intensity of radiation received by the samples when placed in the exposure area [6, 7]. Therefore it is necessary to qualify the photostability chamber in order to prove that this equipment is definitely suitable to realize photostability testing, presenting the characteristics of operationalization and performance capable of ensuring the reliability of the results provided by it. Basicly two fundamental aspects are relevant in photostability studies:

1. the irradiation source and
2. sample exposure or dose.

In the next sub-Section the main sources of radiation used in photostability studies will be described and also how to estimate the radiation dose used upon the sample. Some photochemical and photophysical concepts will be presented as they are needed. One of the current problems found in the photostability field is the understanding of the used terminologies. For instance, some terms found frequently in commercial literature applied properly in color matching and lighting purposes are inappropriate for pharmaceutical photostability testing purposes.

Irradiation Sources

Many different irradiation sources can be used in the photostability studies of drugs and drug products. The sources used should be comparable in SPD to those to which products are exposed in practical use. The SPD is the energy level of a light source through a range of wavelengths of light (spectrum), in other words, is represents the true "fingerprint" of the light source. It is however, difficult to predict the actual exposure of a pharmaceutical product during practical usage. Of all of the topics in the International Conference on Harmonisation (ICH) Q1B document, none is more important than that related to the source. The source is the "active agent" in photostability studies. Without control of the source, results are questionable [8]

In terms of the possibility of photochemical reaction induced during the lifetime of the pharmaceutical, the UV component of sunlight is the most potentially damaging. However, there may be long periods of exposure to fluorescent and incandescent lighting during the various stages of manufacture, storage and use, so it is important to consider their spectral distribution as well [9].

The sun is the most important source of ultraviolet (UV) radiation, although visible light and infrared radiation constitute the major fraction of solar EMR reaching the atmosphere of the Earth (Reference, Year). About 8% of the radiation energy reaching the Earth's atmosphere is within the UV spectrum. There are differences in the distribution of the UV light throughout the world due to the scattering and absorption of UV radiation and than depends on several factors: latitude, zenith angle, cloud-cover and thickness of the ozone layer (Reference, Year). Ultraviolet radiation is divided into three different regions: UVC radiation to wavelengths between 200 and 280 nm, UVB radiation between 280 and 320 nm, and UVA between 320 and 400 nm. The UVA is the most abundant of the UV radiation, and UVC is practically absent, as it is absorbed in the atmosphere [11, 12].

Incandescent lamps, halogen lamps and fluorescent tubes are the most commonly used light sources in homes and work places. Artificial lights are also sources of UV radiation. Incandescent lamps give practically no UV radiation, while halogen lamps operate at a higher_temperature and give small fluency rates of UVA radiation. The same is true for common fluorescent tubes. These tubes contain mercury vapour, which gives a typical line spectrum when current is passed through it. The strongest lines are located at 254 nm, 313 nm, 405 nm, 436 nm, 546 nm and 577 nm. The UV lines are partly filtered out by the glass of the tubes and partly converted to visible radiation with a broad spectrum by a phosphorus material layer on the inner surface of the tubes.

Table 1. Summary of ICH radiation sources used in photostability testing

Option	Description	Lamp type	Additional requirements
1	D65/ID65	Not identified	N/A
1	Artificial daylight	Fluorescent	N/A
1	D65 metal halide	Shot-arc	Filter to remove >320 nm radiation
1	D65 Xenon	Short-arc Long-arc	Filter to remove >320 nm radiation
1	ID65 Cool-white	Fluorescent Fluorescent	ISO 10977:1993(E) ISO 10977:1993(E)
2	Near ultraviolet	Fluorescent	Bandwidth: 320–400 nm; λmax = 350–370 nm; 2 significant proportion of UV between 320–360 nm and 360–400 nm

Reference: J.T. Piechock; K. Thoma; Pharmaceutical Photostability and Stabilization Technology, Informa Healthcare, New York, 2007, p.100.

Some of the tubes that are intended to produce pure visible light leak small yields of the mercury lines in the UV region. Different types of the phosphorus material layer result in different emission spectra. All these features show that each lamp chosen to perform a photostability testing must be thoroughly evaluated. Five options are offered in the ICH document (described in Table 1). Among these types, the fluorescent lamps have been very popular since their initial introduction into the market.

Relative spectral intensity curves for sunlight and different lamps are shown in Figure 2. Each of these spectra extends from near 300 nm in the ultraviolet region to beyond 3000 nm in the infrared, but with different intensity distribution.

As described in the ICH guide, a sample is exposed to a source of irradiation and a flux of irradiation reaches it. Flux is the basic unit of optical power and is expressed in two ways:

- *radiant flux* (or radiant power), the total rate of electromagnetic radiation emitted from a source and is a radiometric unit for reporting radiation intensities in terms of energy, expressed in watts (Figure 3);
- *luminous flux*, which is analogous to radiant flux, except that is used to measure the intensity of visible radiation as viewed by responsiveness of the human eye, and are expressed in lumens [Thatcher, 2001] (Figure 4).

The flux reaches all the area around the sample, forming a flux density, or flux per unit area (Figure 3) and is expressed in:

- *irradiance,* the total rate of energy of all wavelengths of the electromagnetic spectrum that is incident upon a given surface area (radiant flux per unit area) and commonly expressed as watts per square meter (W/m^2) (Figure 5);
- *illuminance*, the luminous flux per unit area and pertains only to the visible region of the spectrum. The units of illuminance are lumens per square meter, or lux. An older unit for illuminance is the foot-candle (lumens per square foot). One foot-candle corresponds to 10.76 lux.

Irradiance and illuminance measurements are inversely proportional to the square of the distance from the source. This is known as the inverse square law (energy [E] = irradiance [I] / distance^2 [d^2]); however, this applies only when the ratio of distance to lamp diameter is greater than 5:1.

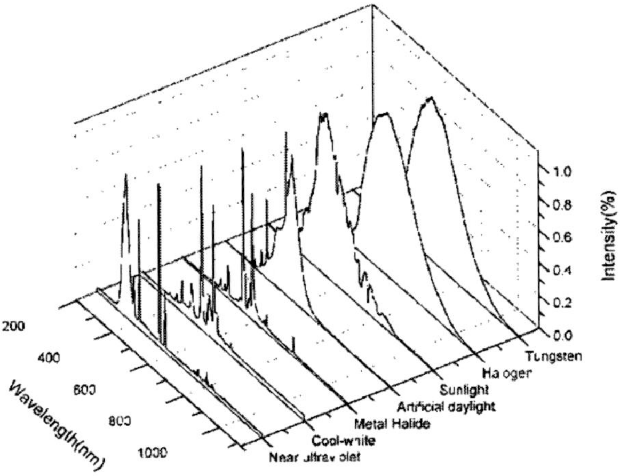

Figure 2. Spectral power distribution plots of the sunlight and of artificial light sources commonly used in indoor lighting.

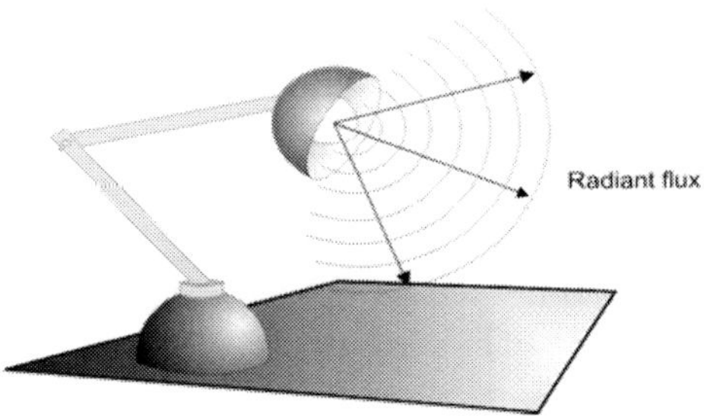

Figure 3. Scheme of the radiant flux.

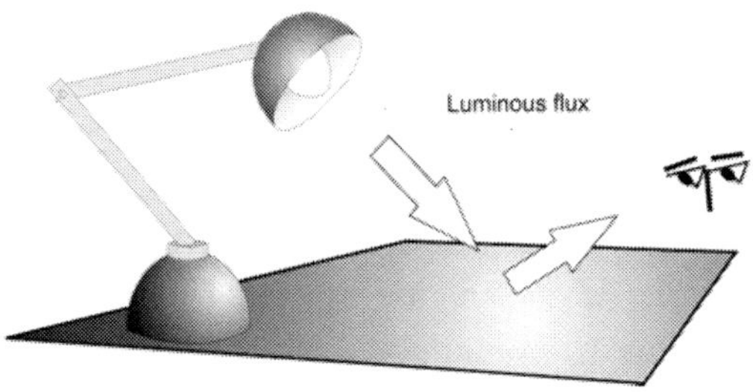

Figure 4. Schematic representation of luminous flux.

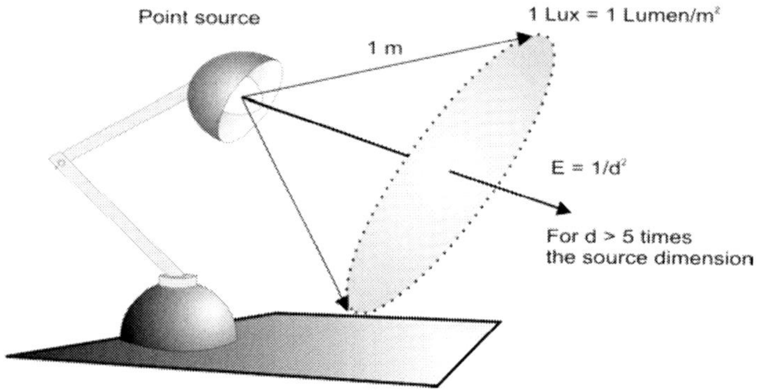

Figure 5. Schematic drawings describing some terms commonly applied in photostability studies.

The Sample Exposure

Photostability testing is typically performed under controlled conditions, often in a sealed chamber where exact exposure levels to the spectra of light can be delivered for precise analysis of the effects. The light levels used in photostability testing are generally high enough to accelerate hours, days, weeks, months, or even years worth of light exposure down to seconds, minutes, or hours in the testing chamber. Monitoring of the exposure levels is critical and is either done by a built in measurement equipment within the chamber or by external instrumentation. This type of exact, accelerated, laboratory level photostability testing is typical for the pharmaceutical, paint, ink, and dye manufacturing industries, among others.

There are two experimental ways to qualify the source of radiation: (i) by applying chemical actinometer systems (chemical compounds with a linear photodegradation rate) or (ii) by using a physical device on which the number of photons applied can be fully determined [6, 7, 13].

The sample exposure, also known as dose, is calculated by measuring the irradiance or the illuminance of a source over a specified amount of time. Irradiation is irradiance multiplied by the time in the near-UV region (320–400 nm). For example, if a sample is exposed to an irradiance of 20 W/m^2 for 10 h the near-UV dose imparted on that sample is 200 W•h/m^2. The same way, the dose calculation in the visible region (400-700 nm), called illumination, is described as illuminance (lux) multiplied by the time and the unit of exposure is lux•h.

Table 2. Brief summary of the most common technical terms used in photophysics [9]

Spectral Region	Radiation Source		Radiation incident on sample surface		Total exposure (dose)	
	Term	Units	Term	Units	Term	Units
Near UV*	Radiant flux	Watts	Irradiance	W/m^2	Irradiation	W•h/m^2
Visible	Luminous flux	Lumens	Illuminance	Lumens/m^2 or lux	Irradiation	lux•h

*200-400 nm.

The irradiance (in the near-UV region) and illuminance (in the visible region) emitted from the source must be determined so that the time of exposure required to meet the minimum guideline exposure limits can be calculated. Table 2 provides a brief summary of terms from source to sample. For general terms the reader should consult other established references such as "The Vocabulary of Photochemistry" by Pitts et al. [14], "Basic terminology of photochemistry" by Thatcher et al. [9], the "Glossary of Terms in Photocatalysis and Radiocatalysis" by Parmon et al. [15] and "Glossary of Terms Used in Photochemistry" by Kuhn, Braslavsky and Schmidt [16].

SPD can be used also for calculating dose for a source in the near-UV and visible regions. The plot of wavelength *versus* calibrated intensity across a spectral range (e.g. 320–800 nm) is supplied by manufacturers most likely will be an average of several sources and not an individual source. The change of SPD depends of the type of source and the frequency of use.

To determine the exposure surface it is necessary to focus on three key-points: (i) confirmation of the adequacy of the emission spectrum of the lamps used by the experimental measurement, (ii) evaluation of illuminance of visible light and (iii) irradiance of ultraviolet radiation provided by the chamber for further photostability testing of photobleaching of drugs. For the last two points it is possible to use physical photocounting devices based on the radiometry. These devices convert electromagnetic energy into electrical current, consisting of a detector with visible or UV filter that allows the passing of defined wavelengths for the sensor. For measurements of visible light, it is recommended the use of luximeters, devices that measure the intensity of visible light perceived by the human eye, in other words, measure the intensity of illumination or illuminance, so it is not sensitive to the other types of emission source. In the other hand radiometers measure the energy intensity of ultraviolet radiation, or irradiance of the UV source [6, 7 10, 13, 17].

The following example is used to show how to qualify the exposure area in a commercial or homemade photostability chamber. In the exemplified case, a homemade chamber was build using three fluorescent lamps, a pair of cool white lamps 5200 lumens each, for emission in the visible spectrum, and a third lamp with ultraviolet spectrum emission (UVA) of 20 W UV power. The lamps were selected in order to agree with the specifications determined by ICH (option 2). The equipment used in the readings of illumination (in lux) was a luximeter model with maximum range 100.000 lux. In order to determine the UV irradiance with reliability two different techniques were used. The first method consisted in the theoretical estimative of the direct incidence of light radiation, based on data of the power emission UV, supplied by the manufacturer of the lamp used. The second method consisted of successive readings of irradiance using a radiometer with maximum range 20.000 $\mu W/cm^2$ on the same grid configuration previously used for the determination of illuminance. The grid is a framework of crisscrossed or parallel bars (a grating or mesh) region that supposedly will receive a greater light intensity and / or power density, homogeneously. The results of the measurements of irradiance and illuminance of the UV and cool white lamps were used to build up graphics of exposure surface (intensity maps) [18, 19]. The emission spectra of each type of lamp used in the photostability chamber and the spectra can be seen in Figure 6. Full line and dashed line show the spectra of UV lamp and cool white lamp, respectively.

In Figure 6, the emission spectra of the lamps confirm the manufacturer´s description. The UV lamp has predominantly emission between 320 and 400 nm, which corresponds exactly to the UVA range. The cool white lamp, despite depicting a more irregular spectrum with several peaks, shows the expected range for a white fluorescent lamp. The uniformity in the visible spectrum is not critical and the radiation in this region has not great impact on the photostability testing, although it is reported that some drugs photobleach when exposed to visible light [7, 17, 20, 21].

The surface exposure maps of the lamps in the photostability chamber can be seen in Figure 7. A non-uniform distribution in the intensity of UV and visible light on the surface of exposure due to the central location of the lamps can be observed. The increase of the distance from the center decreases, as expected, the intensity of light. It is interesting to observe the difference in uniformity of distribution of UV and visible light, even with the chamber's walls covered with aluminum sheets to provide better light distribution.

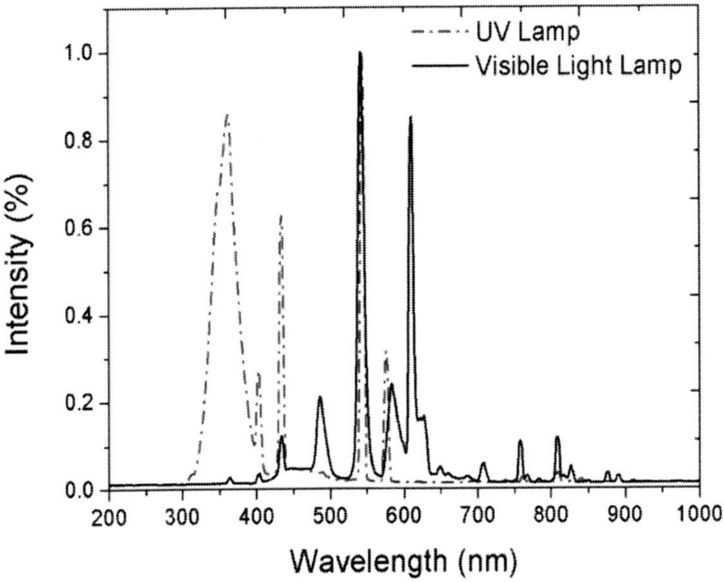

Figure 6. Overlap spectra of ultraviolet lamp (Cleo Performance 100W-S – Phillips) and visible light lamp (Cool White L58W LUMILUX OSRAM) [18, 19].

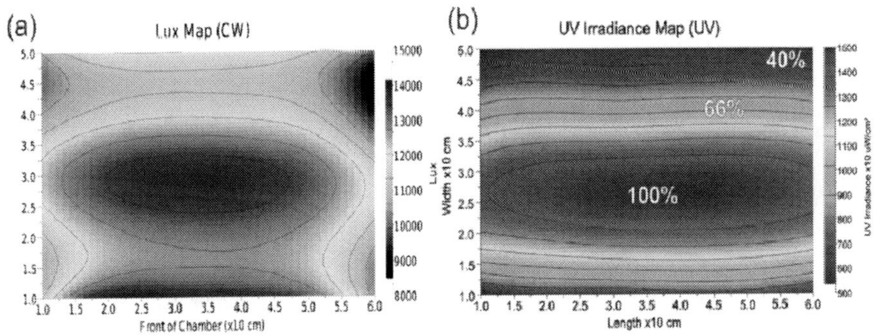

Figure 7. Intensity map on exposure surface of (a) visible light (lux); (b) UV light (uW/cm²) [18, 19].

This distribution profile shows that only a restricted area may be used to induce an uniform photodegradation of the sample. As each photostability chamber has its own design one has to estimate the working area which will guarantee an uniform light output to the drug thus providing a homogeneous degradation kinetics.

ACTINOMETRY

The determination of dose and exposure time is achieved by actinometer systems. According to the "Glossary of terms used in photochemistry" (IUPAC Recommendations, 1996) an actinometer is a chemical system or physical device on which the number of photons in a beam absorbed in a defined space can be fully determined. The actinometry determines the photon flux or "dose of light" exposed on the pharmaceutical products during photostability testing with the use of a chemical method [7, 13, 16, 22, 23].

Although there is an extensive list of substances applied as chemical actinometers (Table 3) there are no well-established actinometric systems sensitive only to UV radiation.

Table 3. Table showing the main actinometer systems described in the literature [16]

Actinometer	Phase	Spetrum range
Uracil	Solid phase	250–400 nm
Nalidixic acid	Solid phase	280–360 nm
Thymine	SDS microemulsions	290–320 nm
o-Nitrobenzaldehyde	Solid phase	280–410 nm
Quinonaphthalone	Solid phase	366–436 nm
Nitrosyl chloride	Gas phase	230–630 nm
Azomethane	Gas phase	270–410 nm
Chlorine	Gas phase	280–380 nm
Acetone	Gas phase	250–320 nm
Potassium reineckaet	Liquid phase	316–750 nm
Azobenzene	Liquid phase	230–460 nm
p-Benzoquinone	Liquid phase	260–380 nm
Phenylglyoxylic acid	Liquid phase	254–405 nm
Hydrogen peroxide	Liquid phase	250–450 nm
Potassium ferrioxalate	Liquid phase	250–500 nm
2,3-Dimethylbut-2-ene	Liquid phase	280–560 nm
o-Nitrobenzaldehyde	Liquid phase	300–410 nm
Benzophenone/benzhydrol	Liquid phase	300–390 nm
Riboflavin	Liquid phase	350–450 nm
Cobalt (III) amines	Liquid phase	488 nm

The known paper of Yoshioka et al. [22] proposed quinine actinometry as an universal standardized method for calibrating UV intensity in light sources and showed that there is a linear correlation between the quinine absorbance and the integrated UV radiation.

The Japanese National Institute of Health Sciences and the Japanese Pharmaceutical Manufacturer's Association suggested its use and later on, the Food and Drug Administration (FDA) adopted this dose measurement as a standard chemical actinometer [4, 35; 22]. The linearity of degradation with UV dose is exemplified in Figure 8 (a) and (b). This substance is chosen as a standard actinometer for UV light also due to its good solubility and relatively low cost. Two options of experimental procedures for monitoring sample exposure to UV fluorescent lamp (based on FDA/National Institute of Standards and Technology study) are described in the guideline ICH Q1B, using an aqueous solution of quinine monohydrochloride dehydrate at 2 % weight/volume.

Option 1 recomends the use of 10 mL of the 2% quinine solution in a 20 mL colorless glass ampoule, hermetically sealed. This is the light probing sample. Another filled ampoule should be used as a control in the chamber, completely wrapped in aluminum foil to fully protect from light. The sample and the control should be exposed to an amount of energy launched from the source for an appropriate time period.

After exposure it is necessary to determine the absorbances of the sample (A_S) and the control (A_0) at 400 nm and to calculate the change in absorbance $\Delta A = A_S - A_0$. The amount of exposure should be sufficient to ensure a change in absorbance of at least 0.9 units. Figure 8(a) shows how the expected change in the absorption profile of the quinine solution under continuous UV radiation exposure. The linearity of the quinine response with the applied dose can be seen in Figure 8(b).

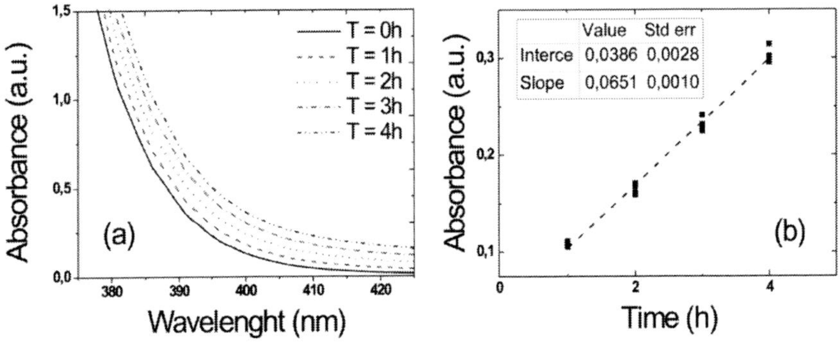

Figure 8. Monitoring the quinine degradation using electronic absorbance spectroscopy (a). The regression curve shows a great linearity for this method (b).

For the option 2, the same procedure should be adopted, but the colorless ampoule must be changed to a quartz cell and kept under radiation in the chamber at enough time to ensure a change in absorbance of at least $\Delta A = 0.5$.

Although quinine actinometry is being applied as a universal standardized method for calibrating UV intensity of light sources since the 90´s many different authors have been arising some questions since the publication of the ICH guidelines. For instance it has been reported by Baertschi et al. (2010) that this method is not only poorly described in the guide but also that its reproducibility depends on the experimental conditions, such as measurement time interval, pH, temperature, oxygenation and lamp emission spectrum [9, 23, 24]. There are some reports that even mention a continuous increase of the quinine absorbance after stopping the radiation source [23]. In the work of Azevedo Filho et al. (2011) [19] tests using quinine and physical actinometry were performed and compared to the results which are used as the basis of the ICH guidelines. The results pointed out that the calibration curve of the quinine solution depends on its concentration and on its location in the chamber and the authors suggested that the quinine actinometry method should not be generalized to any photostability chamber. Moreover in the same work it was suggest that to ensure the reproducibility of the photostability testing and to guarantee a reliable comparison between different chambers, a surface exposure mapping should always be performed and this measurement should be recommended by the ICH guidelines.

Performing a Photostability Testing According to the ICH

Photostability studies are performed for two main purposes: (i) confirmatory studies and (ii) forced degradation.

For confirmatory purposes, according to guideline ICH Q1B, samples should be exposed to light providing an overall illumination of not less than 1.2 million lux•h and an integrated ultraviolet energy of not less than 200 Watt•h/m^2 to allow direct comparisons to be made between the drug substance and the drug product [2, 4, 5].

As mentioned before, ambient light condition is difficult to define since it is dependent on the storage conditions. To get a rough idea of this exposure, for example, consider a 20 W output radiation source at a distance of 1 m from a sample, exposing light continuously 24 h per day. The overall exposure would then be:

$$Exposure = 20\ Watts\ x\ (365\ days\ year\ x\ 24\ h/\ day)/\ 2\ \pi\ (1\ m)^2 = 275.064\ W{\bullet}h/m^2.$$

UV radiation represents only a fraction of the total radiation that is released from light sources, and this includes direct and indirect (indoors) sunlight, fluorescent lights and incandescent lamps. For this reason, the ICH standard for UV exposure is 200 Wh/m^2, whereas for visible light, following the same reasoning and using the above approximation, a year's light exposure would then equal 48 millions lux•h, however, since be fraction of the light likely to cause harm is significantly less then this (one lux is equivalent to 1.46 mW/m^2 at 555 nm), the standard exposure requirement set by ICH are 1.2 millions lux•h. It is possible to achieve the ICH exposure conditions using photostability chambers for both ultraviolet and visible exposure is less then a week [5].

For forced degradation testing, the drug substance alone (powder, solution or suspension) should be placed in chemically inert and transparent containers and a variety of exposure conditions may be used, depending on the photosensitivity of the drug substance involved and the intensity of the light sources. Under forced conditions, by-products arising from the decomposition process may be observed [25-29]. This feature is unlikely to occur in confirmatory studies enabling the use of this information to develop a suitable analytical method.

A number of different lamps have been used in pharmaceutical photodegradation testing [4, 6]. Often the desired exposure can be achieved by a combination of an artificial daylight fluorescent lamp with an ultraviolet emitting lamp. According to the ICH two options of light sources may be used for photostability testing, however for both options it is necessary maintain an appropriate control of temperature to minimize the thermal effect. To monitore this effect it should be included a dark control together with the analized sample [4].

The first option of light source is designed to produce an output similar to the D65 (standard for outdoor daylight) or ID65 (indoor indirect daylight standard) emission such as an artificial daylight fluorescent lamp, xenon, or metal halide lamp (SPD´s are shown in Figure 2). If the source is shown to emit significant radiation below 320 nm, an appropriate filter may be included to eliminate such radiation [4].

For option 2 the same sample should be exposed to independents lamps, a cool white fluorescent and an ultraviolet lamp. The spectral distribution of the cool white fluorescent lamp should be from 400 to 700 nm, and a UV fluorescent lamp from 320 to 400 nm, with a maximum energy emission between 350 and 370 nm [4].

Although complying with the guideline demands the choice of the irradiation method may affect test results. High irradiances may shorten testing times, but can lead to high yields of photodegradation. Many different examples of photostability testing of diverse pharmaceutical products in liquid and in solid form have been shown in the literature [25-34].

SUMMARY

Light can have effects on the active principle in a drug formulation, as well as on the final product or packaging. This may be observed as bleaching of colored compounds or a discoloration of colorless products.

Visible light and UVA are the most relevant spectral ranges of concern due to the abundance of both types in sunlight and typical indoor lighting which are the most likely light sources to be encountered by these photo sensitive products.

The present Chapter focuses on the photostability testing commonly performed for pharmaceutical products, its main steps, conditions, difficulties and current updates. It is emphasized that measurement of the light source exposure levels on the object is crucial to obtain reliable results that are used to predict the product behavior upon light exposure.

ACKNOWLEDGMENTS

The authors are grateful to Prof. Adriana Fontes (UFPE), Prof. José Almiro da Paixão (UFPE) and Steven W. Baertschi (Eli Lilly and Company) for helpful discussions; to Pedro Oliveira for helping with the drawings and the financial support of the following agencies: FACEPE, CAPES, PROPESQ/UFPE and CNPQ.

References

[1] ICH. Guideline Q1A(R2): Stability Testing of New Drug Substances and Products (Second Revision). *Federal Register*, 225 (2003) 65717-18.

[2] Lachman, L., Lieberman, H.A., Kanig, J.L.; Teoria e Prática na Indústria Farmacêutica, *(Translated from the original: The Theory and Pratice of Industrial Pharmacy)*, Fundação Calouste Gulbenkian, Lisboa (2001), p.1277 – 1288.

[3] WORLD HEALTH ORGANIZATION, International Stability Testing: guidelines for stability testing of pharmaceutical products containing well established drug substances in conventional dosage forms. Annex 5, *WHO Technical Report Series*. 863, 1996.

[4] ICH. Guidance for Industry Q1B Photostability Testing of New Drug Substances and Products. *Federal Register*, 62 (1996) 27115 – 27122.

[5] Waterman, K.C.; Adami, R.C.; Accelerated aging: Prediction of chemical stability of pharmaceuticals. *International Journal of Pharmaceuticals,* 293 (2005) 101-125.

[6] Reynolds, D.W.; Facchine, K.L.; Mullaney, J.F.; Alsante, K.M.; Hatajik, T.D.; Motto, M.G.; Available Guidance and Best Practices for Conducting Forced Degradation Studies. *Pharmaceutical Technology, February* (2002) 48-56.

[7] Piechock, J.T.; Thoma, K.; Pharmaceutical Photostability and Stabilization Technology, Informa Healthcare, New York, 2007.

[8] Tønnesen, H.H.; *Photostability of Drugs and Drug Formulations.* second ed., CRC Press LLC, New York, 2004.

[9] Templeton, A.C.; Xu, H.; Placek, J.; Reed, R.A.; Implications of Photostability on the Manufacturing, Packaging, Storage, and Testing of Formulated Pharmaceutical Products. *Pharmaceutical Technology, March* (2005) 68-86.

[10] Thatcher, S.R., Mansfield, R.K., Miller, R.B., Davis, C.W., Baertschi, S.W., Pharmaceutical photostability: a technical guide and practical interpretation of the ICH guideline and its application to pharmaceutical stability – Part I and II, *Pharm. Technol.* 25 (2001) 58–60.

[11] Tønnesen, H.H.; Formulation and stability testing of photolabile drugs; *International Journal of Pharmaceutics*, 225 (2001) 1-14.

[12] Madronich, S., Mckenzie, R.L., Bjorn, L.O., Cadwell, M.M.; Changes in biologically active ultraviolet radiation reaching the Earth's surface. *Journal of Photochemistry and Photobiology* B: Biology 46 (1998) 5-19.

[13] Moan, J.: Visible light and UV radiation. In: Radiation at home, outdoors and in the workplace. Editors: D. Brune, R. Hellborg B. R. R. Person and R. Pääkkönen, pp 69–85. *Scandinavian Science Publisher*, Oslo (2001).

[14] Allen, J.M.; Allen, S.K.; Baertschi, S.W.; 2-Nitrobenzaldehyde: a convenient UV-A and UV-B chemical actinometer for drug photostability testing. *Journal of Pharmaceutical and Biomedical Analysis*, 24 (2000) 167-178.

[15] Pitts, J.N. Jr. , Wilkinson, F. , Hammond, G.S. The vocabulary of photochemistry. *Advances in Photochemistry* . New York : Wiley Interscience , 1963 .

[16] Parmon, V. , Emeline, A.V., Serpone, N. . Glossary of terms in photocatalysis and radiocatalysis. Int J Photoenerg 2002 ; 4 : 91 – 131.

[17] Kuhn, H.J., Braslavsky, S.E., Schmidt, R.; Chemical actinometry, *Pure Appl. Chem.* 76 (2004) 2105–2146.

[18] Moore, D.E.; Mechanisms of photosensitization by phototoxic drugs; *Mutation Research* 422 (1998) 165–173.

[19] Azevedo Filho C.A.; Construção de uma câmara de fotoestabilidade, estudo da fotodegradação da vitamina A e da ação antioxidante do polifosfato de sódio em misturas de nutrição parenteral, Dissertação (Mestrado em Ciências Farmacêuticas) – Departamento de Ciências Farmacêuticas, *Universidade Federal de Pernambuco*, Recife, 2008.

[20] Azevedo Filho, C.A., Gomes, D.F., Guedes, J.P.M., Batista, R.M.F., Santos, B.S.; Considerations on the quinine actinometry calibration method used in photostability testing of pharmaceuticals. *Journal of Pharmaceutical and Biomedical Analisys*, 54 (2011) 886–888.

[21] Glass, B. D., Novák, Cs.; Brown, M. E.; The thermal and photostability of solid pharmaceuticals. *Journal of Thermal Analysis and Calorimetry*, 77 (2004) 1013–1036.

[22] Schnabel, W.; *Polymers and Light Fundamentals and Technical Applications*. WILEY-VCH Verlag Gmb HandCo. KGaA (2007), Weinheim, Germany.

[23] Yoshioka, S.; Ishihara, Y.; Terazono, T.; Tsunakawa, N.; Murai, M.; Yasuda T.; Y.; Kitamura , Kunihero, Y.; Sakai, K.; Hirose, Y.; Tonooka, K.; Takayama, K.; Imai, F.; Godo, M; Matsuo, M; Nakamura, K.; Aso, Y.; Kojima, S.; Takeda, Y.; Terao, T; Quinine Actinometry as a Method for Calibrating Ultraviolet Radiation Intensity in Light- stability Testing of Pharmaceuticals, *Drug Development and Industrial Pharmacy*, 20 (1994) 2049-2062.

[24] Baertschi, S.W.; Commentary on the quinine actinometry system described in the ICH draft guideline on photostability testing of new drug substances and products. *Drug Stability* 1 (1997), pp. 193 – 195.

[25] Baertschi, S.W., Alsante, K.M., Tønnesen, H.H.; A critical assessment of the ICH guideline on photostability testing of new drug substances and products (Q1B): recommendation for revision, *J. Pharm. Sci.* 99 (2010), pp. 2934–2940.

[26] Nudelman, N.S., Cabrera, C.G.; Spectrofluorimetric assay for the photodegradation products of alprazolam. *Journal of Pharmaceutical and Biomedical Analysis,* 30 (2002) 887-893.

[27] Garcia, C.V., Nudelman, N.S., Steppea, M, Schapovala, E.E.S.; Structural elucidation of rabeprazole sodium photodegradation products. *Journal of Pharmaceutical and Biomedical Analysis*, 46 (2008) 88-93.

[28] Huidobro, A.L., Rupérez, F.J. , Barbas, C.; Isolation, identification and determination of the major degradation product in alprazolam tablets during their stability assay. *Journal of Pharmaceutical and Biomedical Analysis*, 44 (2007) 404–413.

[29] Kochanskya, J., Nasrb, M.; Laboratory studies on the photostability of fumagillin, *the active ingredient of Fumidil* B; Apidologie, 35 (2004) 301-310.

[30] Mancini, F., Bolognesi, M.L., Melchiorre, C., Andrisano V.; Investigation of the photostability properties of memoquin, a quinone derivative for the treatment of Alzheimer's disease. *Journal of Pharmaceutical and Biomedical Analysis,* 50 (2009) p.164–170.

[31] Aman, W., Thoma, K.; How to Photostabilize Molsidomine Tablets; *Journal of Pharmaceutical Sciences*; 93(7), 2004.

[32] Javidnia, K., Miri, R.; Movahedb, L., Golrangi, S.; Photostability Determination of Commercially Available Nifedipine Oral Dosage Forms in Iran. *Iranian Journal of Pharmaceutical Research*, 2 (2003) 111-115.

[33] Mancinia, F., Bolognesia, M.L., Melchiorrea, C., V. Andrisano; Investigation of the photostability properties of memoquin, a quinone derivative for the treatment of Alzheimer's disease. *Journal of Pharmaceutical and Biomedical Analysis,* 50 (2009) 164-170.

[34] Ioele, G., Cione, E., Risoli, A., Genchi, G., Ragno, G.; Accelerated photostability study of tretinoin and isotretinoin in liposome formulations. *International Journal of Pharmaceutics*, 293(2005) 251-260.

[35] Kojima, T., Onoue, S., Katoh, F., Teraoka, R., Matsuda, Y., Kitagawa, S., Tsuhako, M.; Effect of spectroscopic properties on photostability of tamoxifen citrate polymorphs. *International Journal of Pharmaceutics*, 336 (2007) 346-351.

[36] Drew, H.D., L.K. Thornton, W.E. Juhl, and J.F. Brower, "An FDA/PhRMA Interlaboratory Study of the International Conference on Harmonization's Proposed Photostability Testing Procedures and Guidelines," *Pharmacopeial Forum*, 24(3):6317, 1998.

INDEX

A

B

D

E

F

G

H

M

Q

R

S